STRUCTURE AND BONDING

Volume 19

Editors: J. D. Dunitz, Zürich
P. Hemmerich, Konstanz · R. H. Holm, Cambridge
J. A. Ibers, Evanston · C. K. Jørgensen, Genève
J. B. Neilands, Berkeley · D. Reinen, Marburg
R. J. P. Williams, Oxford

With 37 Figures

Springer-Verlag
Berlin Heidelberg GmbH 1974

ISBN 978-3-662-15540-0   ISBN 978-3-540-37850-1 (eBook)
DOI 10.1007/978-3-540-37850-1

Originally published by Springer-Verlag Berlin Heidelberg New York in 1974.

Softcover reprint of the hardcover 1 st edition 1974

Library of Congress Catalog Card Number 67-11280.

Contents

STRUCTURE AND BONDING is issued at irregular intervals, according to the material received. With the acceptance for publication of a manuscript, copyright of all countries is vested exclusively in the publisher. Only papers not previously published elsewhere should be submitted. Likewise, the author guarantees against subsequent publication elsewhere. The text should be as clear and concise as possible, the manuscript written on one side of the paper only. Illustrations should be limited to those actually necessary.

SPRINGER-VERLAG		SPRINGER-VERLAG NEW YORK INC.
D-6900 Heidelberg 1	D-1000 Berlin 33	
P. O. Box 105280	Heidelberger Platz 3	175, Fifth Avenue
Telephone (06221) 487·1	Telephone (030) 822001	New York, N. Y. 10010
Telex 04-61723	Telex 01-83319	Telephone 673-2660

Relationships between Covalency, Interatomic Distances, and Magnetic Properties in Halides and Chalcogenides

R. D. Shannon*

Central Research Department E. I. du Pont de Nemours & Co. Wilmington. Del., U.S.A.

H. Vincent

Laboratoire des Rayons X — CNRS B. P. N⁰ 166 — Centre de Tri 38042 — Grenoble Cedex, France

Table of Contents

Abstract

A "covalency contraction" parameter, R_V, is defined as the ratio of the unit cell volume of a transition metal compound M_mX_n relative to the unit cell volume of Mg_mX_n. This parameter is found to be proportional to the electronegativity of X and thus inversely proportional to the degree of covalence of the M—X bond. The parameter R_V explains the relative

* Contribution No. 2131.

differences in cell volumes (V) for certain pairs of isotopic compounds, *e.g.*, V(FeO) is larger than V(MgO), but $V(Fe_2GeS_4)$ is smaller than $V(Mg_2GeS_4)$. Similar but smaller variations in R_V occur in ternary oxides $M_xY_yO_z$ and can be traced to the covalence of the Y—O bond. Approximately linear relationships exist between R_V and spin-transfer coefficients determined for Ni^{2+}, Co^{2+} and Mn^{2+} from magnetic resonance and neutron diffraction methods, and between R_V and the Mössbauer isomer shift for Fe^{2+}. The differences in spin transfer observed for MnO and $MnCO_3$ are consistent with the differences in R_V for the two compounds.

I. Introduction

The effective ionic radii of *Shannon* and *Prewitt* (1969) can frequently be used to predict average interatomic distances and to correlate unit cell volumes of series of isostructural oxides and fluorides. However, some systematic discrepancies were recently found in tetrahedral oxy-anion distances and in the unit cell volumes of certain series of fluoride compounds. It was pointed out by *Banks*, *Greenblatt*, and *Post* (1970) that the observed V—O distances in Ca_2VO_4Cl are smaller than those predicted by the effective ionic radii. Subsequently, the discrepancies in Ca_2VO_4Cl and other tetrahedral oxy-anion distances were attributed to covalency effects (*Shannon*, 1971, and *Shannon* and *Calvo*, 1972) in which bonds exhibiting a greater degree of covalency were assumed to shorten.

Anomalies in the relative sizes of Ni^{2+} and Mg^{2+} in octahedral coordination were found by *Longo* and *Kafalas* (1969) in the series of perovskites, $CsMF_3$, and by *Shannon* and *Prewitt* (1970) in the series of rutiles MF_2, where M = Ni, Mg, Co, Fe, and Mn. In both of these isostructural series the unit cell volume of the compound containing Ni was found to be greater than the volume of the compound containing Mg. This is in contradiction to the relative size of Ni^{2+} and Mg^{2+} in oxides.

Similar discrepancies in interatomic distances in Fe_2GeS_4 and Mg_2GeS_4 were recently noted by *Vincent* and *Perrault* (1971) and *Vincent* and *Bertaut* (1972). Both the interatomic distances and the unit cell dimensions of the Fe compounds are smaller than those of the Mg compounds in contradiction to the ionic radii of Fe^{2+} (high-spin) (0.78 Å) and Mg^{2+} (0.72 Å) in oxides. The same behavior had been noted earlier by *Patrie* and *Chevalier* (1966) in the compounds ML_2S_4 where M = Mg, Fe, Cr, and Mn and L = rare earth.

In order to explain these apparent contradictions, we have derived a "covalency contraction" parameter, R_V, which indicates the change of the distance of a covalent bond M—X relative to the distance of the less covalent bond Mg—X. In the first part of this paper, we show how systematic decreases in the ratio, R_V, and thus in the distances M—X accompany changes in the electronegativity differences between M and X. Consequently, we show how varying degrees of covalence of the M—X bond can explain these anomalies in relative unit cell volumes of compounds containing Mg^{2+}, Fe^{2+}, Ni^{2+}, Co^{2+}, Mn^{2+}, Zn^{2+}, Cd^{2+}, Cr^{2+} and In^{3+}. In the second part of the paper, we show a direct relationship (1) between the parameter, R_V, and the reduction in the magnetic moment for compounds containing Mn^{2+}, Co^{2+}, and Ni^{2+} and (2) between the parameter, R_V, and the Mössbauer isomer shift in compounds containing Fe^{2+}, and thus show consistent relationships between covalency, interatomic distance and magnetic properties of the transition metal ions, Mn^{2+}, Fe^{2+}, Co^{2+}, and Ni^{2+}.

II. Effects of Covalency on Interatomic Distances

A. Procedure

In order to see the effect of covalency on M_1—X distances it is necessary to have for comparison a set of M_2—X distances, which are relatively more ionic than M_1—X bonds[1]). We choose two parameters for comparing M—X distances: 1) ratios of the cube of the mean interatomic distances relative to the Mg compound, $R_d = \langle M_1\text{—}X\rangle^3 / \langle Mg\text{—}X\rangle^3$ in compounds having the same structure with differing anions, *e.g.*, X = O, S, and Se and 2) ratios of unit cell volumes $R_V = V(M_mX_n) / V(Mg_mX_n)$ for isotypic compounds with different anions. We call the ratios R_d and R_V "covalency contraction parameters". The use of R_d has the advantage of allowing comparison of actual distances but is not as useful as R_V because of the relative scarcity of structural data. However, it is possible to compare distances from different structures if certain precautions are taken. In order to maintain the same structure for comparison (2) it is also necessary that the effective ionic radius of M_1 be close to that of M_2 *i.e.*, $r_{M1} \sim r_{M2}$. The condition that M_1—X be more covalent than M_2—X implies that $\varkappa_{M1}$ be considerably greater than $\varkappa_{M2}$. Accordingly, there are only a few possibilities for M_2 for any M_1, *e.g.*, Mg^{2+} can be

1) As a measure of covalency we use, for convenience, electronegativities. The scales of *Gordy* and *Thomas* (1956), *Allred* (1961) or *Batsonov* (1968) give approximately equal values; we prefer the Batsonov table because it is more complete.

compared with Mn^{2+}, Fe^{2+}, Co^{2+}, Ni^{2+}, Zn^{2+}, Cd^{2+} and Cr^{2+}; and Sc^{3+} or Y^{3+} with In^{3+}. The unit cell data were obtained from *Wyckoff* (1960), *Donnay* and *Ondik* (1973), and the ASTM card index unless specific references are given. Where a compound contains more than one cation, the ratio of unit cell volumes, R_V, is obviously sensitive to the proportion of M, relative to other cations present[2]). We have corrected this ratio empirically when possible. For example in Mn- and Mg- containing sulfides, we used $R_V = R_V' \pm 0.0002$ (P_M), where P_M = percentage of Mn or Mg relative to all other cations. The value of 0.002 was derived from the series of sulfides listed in Table 1. Similarly, for Co- and Mg-containing fluorides $R_V = R_V' + 0.001$ (P_M). When more than one value of R_V was available we have taken the mean value, $\bar{R}_V$, as characteristic of a particular M—X bond. When insufficient data exist to determine R_V from R_V' we have taken the average of the R_V' values for $\bar{R}_V$. In the case of the oxides, $M_xY_yO_z$, because the value of R_V' was found to depend on the nature of the Y—O bond, we have generally used the value R_V' for the pure oxide MO.

In order to see the effect of covalence on the M_1—X *vs* M_2—X distances, we compare the parameters $\bar{R}_V$ or $\bar{R}_d$ to the electronegativity difference between M_1 and X, $\Delta\varkappa_{M-X}$. Tables 1.A—1.D list unit cell volumes and ratios of V_{M1}/V_{M2} in halides, chalcogenides, and hydroxides for M_1 = Ni, Co, Fe, and Mn; M_2 = Mg; for the pair M_1 = Cd, M_2 = Ca; and for the pair M_1 = In and M_2 = Sc. Tables 2.A—2.E list mean interatomic distances in structures containing Mg, Ni, Co, Fe, Mn, Zn, Cd, In, In, and Sc and the parameter $\bar{R}_d$.

B. Results and Discussion

Table 3 shows that very different unit cell volume ratios, $\bar{R}_V$, result when the anion is changed. We have also compared, when possible, the ratios of the cube of the mean interatomic distances, $\bar{R}_d$. Both ratios show the same dependence on the anion electronegativity. In Figs. 1—7 the parameters $\bar{R}_V$ and $\bar{R}_d$ are plotted *vs* $\Delta\varkappa_{M-X}$. It is evident that the degree of covalence of the M_1—X *vs* the M_2—X bond is important in determining their relative bond lengths. The effect is particularly striking for the pairs Ni—Mg, Co—Mg, Fe—Mg, Mn—Mg, Zn—Mg, and Cr—Mg because the ratio $\bar{R}$ changes from a value >1 to a value <1. This cross-over occurs between F and O for Ni; O and Cl for Co and Fe; S and Se

---

[2]) For the two hypothetical compounds $MgM_{99}O_{100}$ and $MnM_{99}O_{100}$, $\frac{V[Mn]}{V[Mg]} \simeq 1.00$.

Table 1. Unit cell dimensions of isotypic compounds

A. Ni and Mg

Halides				
Compound	V	R'_V	R_V[a]	$\bar{R}_V$
NiF_2	33.4	1.024	1.024	1.066
MgF_2	32.6			
$NaNiF_3$	57.0	1.018	1.07	
$NaMgF_3$	56.0			
$KNiF_3$	64.6	1.030	1.080	
$KMgF_3$	62.7			
$CsNiF_3$	82.7	1.023	1.073	
$CsMgF_3$	80.8			
$BaNiF_4$	348.2	1.001	1.050	
$BaMgF_4$	347.7			
K_2NiF_4	104.9	1.008	1.08	
K_2MgF_4	104.1			
Rb_2NiF_4	114.5	1.010	1.08	
Rb_2MgF_4	113.4			
Tl_2NiF_4	116.7	1.008	1.08	
Tl_2MgF_4	115.8			
Ba_2NiF_6	130.01	1.003	1.070	
Ba_2MgF_6	129.59			
$NiCl_2$	182.8	0.900		0.91
$MgCl_2$	203.2			
$CsNiCl_3$	264.7	0.935		
$CsMgCl_3$	283.8			
$NiBr_2$	72.90	0.925		0.92
$MgBr_2$	78.85			
$CsNiBr_3$	304.0	0.935		
$CsMgBr_3$	325.1			
NiI_2	88.3	0.865		0.86
MgI_2	102.1			

Chalcogenides				
NiO	72.43	0.970		0.97
MgO	74.68			
$Ni_3V_2O_8$	556.6	0.965		
$Mg_3V_2O_8$	576.6			

[a] $R_V = R'_V + .001\ (P_M)$.

Table 1. (continued)

Chalcogenides				
Compound	V	R'_v	R_v	$\bar{R}_v$
$NiWO_4$	255.9	0.976		
$MgWO_4$	262.2			
$NiMoO_4$	254.6	0.973		
$NgMoO_4$	261.7			
$NiCrO_4$	277.6	0.953		
$MgCrO_4$	291.3			
$NiUO_4$	282.2	0.948		
$MgUO_4$	297.7			
NiP_2O_6	104.0	0.979		
MgP_2O_6	106.2			
$NiIn_2S_4$	1145.7	0.939		0.94
$MgIn_2S_4$	1120.6			

Hydroxides				
$Ni(OH)_2$	38.67	0.945		0.95
$Mg(OH)_2$	40.90			
$NiSn(OH)_6$	459.2	0.975		
$MgSn(OH)_6$	470.9			
$NiSiF_6 \cdot 6H_2O$	721.9	0.922		
$MgSiF_6 \cdot 6H_2O$	782.8			

B. Co and Mg

Halides				
Compound	V	R'_v	R_v[b]	$\bar{R}_v$
CoF_2	35.0	1.074	1.074	1.090
MgF_2	32.6			
$CsCoF_3$	82.7	1.023	1.073	
$CsMgF_3$	80.8			
$BaCoF_4$	360.4	1.036	1.086	
$BaMgF_4$	347.7			
K_2CoF_4	108.6	1.043	1.110	
K_2MgF_4	104.1			
Rb_2CoF_4	116.9	1.031	1.100	
Rb_2MgF_4	113.4			

[b]) $R_v = R'_v + .001\ (P_M)$.

Table 1. (continued)

Halides

Compound	V	R'_v	R_v	$\bar{R}_v$
Ba_2CoF_6	132.1	1.019	1.086	
Ba_2MgF_6	129.6			
$CoCl_2$	190.1	0.935		0.93
$MgCl_2$	203.2			
$CoBr_2$	71.97	0.913		0.91
$MgBr_2$	78.85			
CoI_2	90.3	0.884		0.88
MgI_2	102.1			

Chalcogenides

CoO	77.67	1.040	1.040	
MgO	74.68			
Co_2SiO_4	296.2	1.021		
Mg_2SiO_4	290.0			
$CoSiO_3$	425.1	1.023		
$MgSiO_3$	415.5			
$CoMoO_4$	260.2	0.994		
$MgMoO_4$	261.7			
$CoWO_4$	262.3	1.000		
$MgWO_4$	262.2			
$CoCrO_4$	286.1	0.985		
$MgCrO_4$	290.6			
$CoUO_4$	293.4	0.985		
$MgUO_4$	297.7			
CoP_2O_6	107.0	1.007		
MgP_2O_6	106.2			
CoV_2O_6	199.5	0.985		
MgV_2O_6	202.6			

Hydroxides

$Co(OH)_2$	40.45	0.989	0.99	
$Mg(OH)_2$	40.90			
$CoSn(OH)_6$	465.3	0.996		
$MgSn(OH)_6$	467.1			
$CoSiF_6 \cdot 6H_2O$	729.0	0.931		
$MgSiF_6 \cdot 6H_2O$	782.8			

Table 1. (continued)

C. Fe and Mg

Halides				
Compound	V	R'_v	R_v	$\bar{R}_v$
FeF_2	36.5	1.120	1.120	1.116
MgF_2	32.6			
$NaFeF_3$	60.7	1.084	1.13	
$NaMgF_3$	56.0			
$KFeF_3$	69.90	1.115	1.16	
$KMgF_3$	62.71			
$CsFeF_3$	81.3	1.068	1.118	
$CsMgF_3$	76.1			
K_2FeF_4	112.2	1.068	1.14	
K_2MgF_4	104.1			
Ba_2FeF_6	135.5	1.046	1.113	
Ba_2MgF_6	129.6			
$FeCl_2$	196.5	0.967		0.97
$MgCl_2$	203.2			
$FeBr_2$	75.22	0.954		0.95
$MgBr_2$	78.85			
FeI_2	96.02	0.940		0.94
MgI_2	102.12			

Chalcogenides				
FeO	80.79	1.082	1.08	
MgO	74.68			
Fe_2SiO_4	307.9	1.062		
Mg_2SiO_4	290.0			
$FeSiO_3$	438.9	1.056		
$MgSiO_3$	415.5			
$FeWO_4$	267.4	1.020		
$MgWO_4$	262.2			
$FeMoO_4$	264.2	1.010		
$MgMoO_4$	261.7			
Fe_2GeS_4	530.7	0.930	0.924	0.964
Mg_2GeS_4	570.3			
$FeIn_2S_4$	1190.3	0.975	0.962	
$MgIn_2S_4$	1220.6			
$FeYb_2S_4$	1273.0	0.968	0.955	
$MgYb_2S_4$	1315.5			

Table 1. (continued)

Chalcogenides				
Compound	V	R'_v	R_v c)	$\bar{R}_v$
$FeLu_2S_4$	1262.2	0.962	0.949	
$MgLu_2S_4$	1312.6			
$FeSc_2S_4$	1165.9	0.971	0.958	
$MgSc_2S_4$	1200.1			
FeY_4S_7	522.9	0.985	0.969	
MgY_4S_7	530.9			
$FeHo_4S_7$	519.2	0.990	0.974	
$MgHo_4S_7$	524.6			
$FeEr_4S_7$	513.7	0.989	0.973	
$MgEr_4S_7$	519.3			
$FeTm_4S_7$	509.5	0.993	0.977	
$MgTm_4S_7$	512.8			

Hydroxides				
$Fe(OH)_2$	42.33	1.035		1.02
$Mg(OH)_2$	40.90			
$FeSn(OH)_6$	466.7	0.999		
$MgSn(OH)_6$	467.1			

D. Mn and Mg

Halides				
MnF_2	39.3	1.205	1.205	1.173
MgF_2	32.6			
$KMnF_3$	73.3	1.169	1.219	
$KMgF_3$	62.7			
$CsMnF_3$	84.0	1.104	1.154	
$CsMgF_3$	76.1			
$BaMnF_4$	381.4	1.097	1.147	
$BaMgF_4$	347.7			
K_2MnF_4	115.5	1.110	1.18	
K_2MgF_4	104.1			
Rb_2MnF_4	124.1	1.094	1.16	
Rb_2MgF_4	113.4			
Ba_2MnF_6	139.3	1.075	1.14	
Ba_2MgF_6	129.6			

c) $R_v = R'_v - .0002\ (P_M)$.

Table 1. (continued)

Halides				
Compound	V	R_v'	R_v	$\bar{R}_v$
$MnCl_2$ hex	209.8	1.032		1.03
$MgCl_2$	203.2			
$MnBr_2$	78.67	0.998		1.00
$MgBr_2$	78.85			
MnI_2	102.9	1.001		1.00
MgI_2	102.7			

Chalcogenides				
MnO	87.8	1.176		1.17
MgO	74.6			
$MnCO_3$	312.7	1.118		
$MgCO_3$	279.7			
$MnSiO_3$	455.6	1.096		
$MgSiO_3$	415.5			
$MnMoO_4$	275.3	1.051		
$MgMoO_4$	261.7			
$MnUO_4$	313.2	1.052		
$MgUO_4$	297.7			
$MnWO_4$	280.0	1.069		
$MgWO_4$	262.2			
MnP_2O_6	113.5	1.069		
MgP_2O_6	106.2			
MnV_2O_6	205.3	1.013		
MgV_2O_6	202.6			
MnS	142.6	1.014	1.012	1.015
MgS	140.6			
Mn_2GeS_4	575.5	1.009	1.015	
Mg_2GeS_4	570.3			
$MnDy_2S_4$	604.7	1.003	1.017	
$MgDy_2S_4$	603.0			
MnY_2S_4	608.2	1.006	1.020	
MgY_2S_4	604.7			
$MnTb_2S_4$	609.2	1.002	1.016	
$MgTb_2S_4$	607.7			
$MnHo_2S_4$	596.9	1.000	1.014	
$MgHo_2S_4$	596.9			
$MnEr_2S_4$	593.8	1.004	1.018	
$MgEr_2S_4$	590.9			

Table 1. (continued)

Chalcogenides				
Compound	V	R'_v	R_v	$\bar{R}_v$
$MnYb_2S_4$	1312.6	0.998	1.012	
$MgYb_2S_4$	1315.5			
$MnLu_2S_4$	1302.5	0.992	1.006	
$MgLu_2S_4$	1312.6			
$MnSc_2S_4$	1198.8	0.999	1.013	
$MgSc_2S_4$	1200.1			
$MnIn_2S_4$	1223.0	1.002		
$MgIn_2S_4$	1220.6			
MnY_4S_7	528.2	0.995	1.011	
MgY_4S_7	530.9			
$MnDy_4S_7$	531.5	1.001	1.017	
$MgDy_4S_7$	531.1			
$MnHo_4S_7$	525.9	1.002	1.018	
$MgHo_4S_7$	524.6			
$MnEr_4S_7$	520.6	1.002	1.018	
$MgEr_4S_7$	519.3			
$MnTm_4S_7$	514.2	1.003	1.019	
$MgTm_4S_7$	512.8			
MnSe	163.0	1.000	1.000	0.982
MgSe	163.0			
Mn_2SiSe_4	627.0	0.969	0.963	
Mg_2SiSe_4	647.1			
$MnSc_2Se_4$	1371.3	0.997	0.983	
$MgSc_2Se_4$	1375.0			
$MnYb_2Se_4$	1489.3	0.994	0.980	
$MgYb_2Se_4$	1498.8			
$MnLu_2Se_4$	1481.5	0.992	0.978	
$MnLu_2Se_4$	1493.3			
MnTe	96.93	0.747		0.75
MgTe	129.7			

Hydroxides			
$Mn(OH)_2$	45.21	1.105	1.10
$Mg(OH)_2$	40.9		
$MnSn(OH)_6$	491.5	1.052	
$MgSn(OH)_6$	467.1		
$MnSiF_6 \cdot 6H_2O$	794.4	1.015	
$MgSiF_6 \cdot 6H_2O$	782.8		

Table 1. (continued)

E. Zn and Mg

Halides				
Compound	V	R'_v	R_v	$\bar{R}_v$
ZnF_2	34.7	1.064	1.064	1.077
MgF_2	32.6			
$KZnF_3$	66.7	1.063	1.113	
$KMgF_3$	62.7			
$NaZnF_3$	58.3	1.041	1.091	
$NaMgF_3$	56.0			
$CsZnF_3$	78.5	1.031	1.081	
$CsMgF_3$	76.1			
$BaZnF_4$	357.8	1.029	1.079	
$BaMgF_4$	347.7			
K_2ZnF_4	105.4	1.012	1.079	
K_2MgF_4	104.1			
Ba_2ZnF_6	132.19	1.020	1.053	
Ba_2MgF_6	129.59			
$ZnCl_2$ hex	219.1	1.078		1.08
$MgCl_2$	203.2			
ZnI_2 hex	102.3	1.002		1.00
MgI_2	102.1			

Chalcogenides				
ZnO	78.40	1.050	1.05	
MgO	74.68			
$ZnSiO_3$	443.7	1.068		
$MgSiO_3$	415.5			
$ZnMoO_4$	263.2	1.006		
$MgMoO_4$	261.7			
$ZnWO_4$	264.2	1.008		
$MgWO_4$	262.2			
ZnP_2O_6	106.3	1.001		
MgP_2O_6	106.2			
ZnV_2O_6	199.8	0.986		
MgV_2O_6	202.6			
ZnTe hex	110.37	0.851		0.85
MgTe	129.69			

Table 1. (continued)

Hydroxides				
Compound	V	R'_V	R_V	$\bar{R}_V$
$ZnSn(OH)_6$	468.2	0.994		0.99
$MgSn(OH)_6$	467.1			
$ZnSO_4 \cdot 7H_2O$	968.3	0.985		
$MgSO_4 \cdot 7H_2O$	983.5			
$ZnSiF_6 \cdot 6H_2O$	735.9	0.940		
$MgSiF_6 \cdot 6H_2O$	782.8			

F. Cd and Mg

Halides				
$KCdF_3$	81.00	1.29		1.29
$KMgF_3$	62.71			
$CdCl_2$	223.8	1.101		1.10
$MgCl_2$	203.2			
CdI_2	106.4	1.042		1.04
MgI_2	102.1			

Chalcogenides				
CdO	103.5	1.386	1.386	
MgO	74.68			
$CdWO_4$	298.7	1.139		
$MgWO_4$	262.2			
$CdMoO_4$	297.3	1.136		
$MgMoO_4$	261.7			
$CdCrO_4$	343.0	1.177		
$MgCrO_4$	291.3			
$CdIn_2S_4$	1258.6	1.031	1.044	1.023
$MgIn_2S_4$	1220.6			
CdY_4S_7	105.7	1.002	1.018	
MgY_4S_7	105.5			
$CdDy_4S_7$	105.6	1.001	1.017	
$MgDy_4S_7$	105.5			
$CdHo_4S_7$	105.6	1.001	1.017	
$MgHo_4S_7$	105.5			
$CdEr_4S_7$	105.6	1.002	1.018	
$MgEr_4S_7$	105.4			

Table 1. (continued)

Hydroxides				
Compound	V	R'_v	R_v	$\bar{R}_v$
$Cd(OH)_2$	49.76	1.217	1.217	
$Mg(OH)_2$	40.90			
$CdSn(OH)_6$	511.0	1.085		
$MgSn(OH)_6$	470.9			
$CsCdPO_4 \cdot 6H_2O$	517.2	1.035		
$CsMgPO_4 \cdot 6H_2O$	499.8			

G. Cd and Ca

Halides				
CdF_2	156.4	0.959		0.96
CaF_2	163.0			
CdI_2	106.5	0.880		0.88
CaI_2	121.0			

Chalcogenides				
CdO	103.5	0.930		0.94
CaO	111.3			
$CdWO_4$	298.7	0.955		
$CaWO_4$	312.6			
$CdMoO_4$	297.3	0.952		
$CaMoO_4$	312.2			

Hydroxides				
$Cd(OH)_2$	49.76	0.913		0.91
$Ca(OH)_2$	54.48			
$CdSn(OH)_6$	511.0	0.951		
$CaSn(OH)_6$	537.0			

H. In and Sc

Halides				
$Ba_3In_2F_{12}$	1749.3	1.025	1.085	1.101
$Ba_3Sc_2F_{12}$	1705.4			

Table 1. (continued)

Halides				
Compound	V	R'_V	R_V	$\bar{R}_V$
K_3InF_6	5554.6	1.043	1.12	
K_3ScF_6	5322.7			
$(NH_4)_3InF_6$	404.7	1.016	1.10	
$(NH_4)_3ScF_6$	398.0			

Chalcogenides				
In_2O_3	64.74	1.085		1.08
Sc_2O_3	59.64			

Hydroxides				
$In(OH)_3$	508.0	1.037		1.04
$Sc(OH)_3$	489.7			
InOOH	78.24	1.035		
ScOOH	75.74			

I. Cr and Mg

Halides				
CrF_2	38.9	1.193	1.193	1.19
MgF_2	32.6			
$NaCrF_3$	64.0	1.142	1.192	
$NaMgF_3$	56.0			
$CsCrCl_3$	283.8	1.002		1.00
$CsMgCl_3$	283.2			

Chalcogenides				
CrY_2S_4	591.9	0.979	0.966	0.977
MgY_2S_4	604.7			
$CrHo_2S_4$	588.4	0.986	0.973	
$MgHo_2S_4$	596.9			
$CrEr_2S_4$	580.6	0.983	0.970	
$MgEr_2S_4$	590.9			

Table 1. (continued)

Chalcogenides				
Compound	V	R'_v	R_v	$\bar{R}_v$
$CrIn_2S_4$	1187.6	0.973	0.960	
$MgIn_2S_4$	1220.6			
CrY_4S_7	106.0	1.005	0.989	
MgY_4S_7	105.4			
$CrHo_4S_7$	105.8	1.002	0.986	
$MgHo_4S_7$	105.6			
$CrEr_4S_7$	105.6	1.002	0.986	
$MgEr_4S_7$	105.5			
$CrTm_4S_7$	105.5	1.003	0.987	
$MgTm_4S_7$	105.2			

Table 2. Mean interatomic distances in halides and chalcogenides containing Mg, Ni, Co, Fe, Mn, and Zn

A. Ni and Mg

Compound	$<M-X>$	$<M-X>^3$	$\bar{R}_d$	$<\bar{R}_d>$
NiF_2	2.00	8.000	1.012	1.021
MgF_2	1.992	7.904		
$KNiF_3$	2.006	8.072	1.030	
$KMgF_3$	1.986	7.833		
$NiCl_2$	2.426	14.278	0.916	0.92
$KMgCl_3$	2.499	15.578		
$CsMgCl_3$	2.496			
NiO	2.084	9.051	0.970	0.96
MgO	2.105	9.327		
$Ni_3V_2O_8$	2.059	8.729	0.949	
$Mg_3V_2O_8$	2.095	9.195		
Ni_2SiO_4	2.089	9.116	0.958	
Mg_2SiO_4	2.119	9.514		
NiS(NiAs)	2.324	12.552	0.713	0.71
MgS(NaCl)	2.601	17.596		
NiSe(NiAs)	2.427	14.296	0.706	0.71
MgSe(NaCl)	2.725	20.235		

Table 2. (continued)

B. Co and Mg

Compound	<M–X>	<M–X>^{3}	$\bar{R}_d$	<$\bar{R}_d$>
CoF_2	2.04	8.489	1.074	1.075
MgF_2	1.992	7.904		
$KCoF_3$	2.035	8.427	1.076	
$KMgF_3$	1.986	7.833		
$CoCl_2$	2.455	14.796	0.950	0.95
$KMgCl_3$	2.499	15.578		
$CsMgCl_3$	2.496			
CoO	2.133	9.704	1.040	1.028
MgO	2.105	9.327		
Co_2SiO_4	2.130	9.663	1.016	
Mg_2SiO_4	2.119	9.514		
$Co_3V_2O_8$	2.091	9.142	0.994	
$Mg_3V_2O_8$	2.095	9.195		
CoS(NiAs)	2.34	12.813	0.728	0.73
MgS(NaCl)	2.601	17.596		
CoSe(NiAs)	2.404	13.893	0.686	0.69
MgSe(NaCl)	2.725	20.235		

C. Fe and Mg

FeF_2	2.08	8.999	1.139	1.139
MgF_2	1.992	7.904		
$KFeF_3$	2.060	8.742	1.116	
$KMgF_3$	1.986	7.833		
$FeCl_2$	2.497	15.570	1.000	1.00
$KMgCl_3$	2.499	15.578		
$CsMgCl_3$	2.496			
FeO	2.1615	10.099	1.083	1.082
MeO	2.105	9.327		
Fg_2SiO_4	2.175	10.289	1.081	
Mg_2SiO_4	2.119	9.514		

Table 2. (continued)

Compound	$\langle M{-}X \rangle$	$\langle M{-}X \rangle^3$	$\bar{R}_d$	$\langle \bar{R}_d \rangle$
FeS (20° C)	2.505	15.719	0.893	
MgS(NaCl)	2.602	17.596		
FeS[a]	2.419	14.155		
MgS(NaCl)	2.602	17.616	0.804	
Fe_2GeS_4	2.534	16.271	0.939	
Mg_2GeS_4	2.588	17.333		
Fe_2SiS_4	2.549	16.562	0.955	
Mg_2GeS_4	2.588	17.333		
FeSe(NiAs)	2.502	15.662	0.774	
MgSe(NaCl)	2.725	20.235		

D. Mn and Mg

MnF_2	2.12	9.528	1.205	1.187
MgF_2	1.992	7.904		
$KMnF_3$	2.093	9.169	1.170	
$KMgF_3$	1.986	7.833		
$NaMnCl_3$ }	2.552 }	16.552	1.062	1.06
$CsMnCl_3$ }	2.545 }			
$KMgCl_3$ }	2.499 }	15.578		
$CsMgCl_3$ }	2.496 }			
MnO	2.222	10.970	1.176	1.150
MgO	2.105	9.327		
$CaMnSi_2O_6$	2.22	10.941	1.124	
$CaMgSi_2O_6$	2.135	9.731		
MnS	2.612	17.820	1.013	1.022
MgS	2.601	17.596		
Mn_2GeS_4	2.615	17.882	1.032	
Mg_2GeS_4	2.588	17.333		
MnSe	2.724	20.212	0.999	1.00
MgSe	2.725	20.235		

[a] NiAs, 190° C.

Table 2. (continued)

E. Zn and Mg

Compound	$\langle M{-}X \rangle$	$\langle M{-}X \rangle^3$	$\bar{R}_d$	$\langle \bar{R}_d \rangle$
ZnF_2	2.03	8.365	1.058	1.061
MgF_2	1.992	7.904		
$KZnF_3$	2.027	8.328	1.063	
$KMgF_3$	1.986	7.833		
ZnO	2.14	9.800	1.051	1.034
MgO	2.105	9.327		
$Zn_3V_2O_8$	2.112	9.420	1.024	
$Mg_3V_2O_8$	2.095	9.195		
$Zn_3P_2O_8$	2.156	10.021	1.050	
$Mg_3P_2O_8$	2.121	9.541		
$Zn_2P_2O_7$	2.112	9.420	1.010	
$Mg_2P_2O_7$	2.105	9.327		

F. Cd and Mg

Compound	$\langle M{-}X \rangle$	$\langle M{-}X \rangle^3$	$\bar{R}_d$	$\langle \bar{R}_d \rangle$
$KCdF_3$	2.146	9.883	1.262	
$KMgF_3$	1.986	7.833		
CdO	2.348	12.945	1.387	1.33
MgO	2.105	9.327		
$Cd_2P_2O_7$	2.31	12.326	1.317	
$Mg_2P_2O_7$	2.107	9.354		
$CdWO_4$	2.28	11.825	1.295	
$MgWO_4$	2.09	9.129		

G. In and Sc

Compound	$\langle M{-}X \rangle$	$\langle M{-}X \rangle^3$		
β-In_2S_3	2.630	18.19	18.20	
$Bi_2In_4S_9$	2.631	18.21		1.042
$CuScS_2$	2.60	17.576	17.47	
Sc_2S_3	2.59	17.37		

Table 3. Tabulation of $\bar{R}_v$ and $\bar{R}_d$ values

	Ni and Mg			Co and Mg			Fe and Mg			Mn and Mg		
	$\bar{R}_v$	$\bar{R}_d$	$\Delta\varkappa$	$\bar{R}_v$	$\bar{R}_d$	$\Delta\varkappa$	$\bar{R}_v$	$\bar{R}_d$	$\Delta\varkappa$	$\bar{R}_v$	$\bar{R}_d$	$\Delta\varkappa$
F^-	1.066	1.021	2.1	1.090	1.075	2.2	1.116	1.128	2.2	1.173	1.187	2.6
Cl^-	0.91	0.92	1.3	0.93	0.95	1.4	0.97	1.00	1.4	1.03	1.06	1.8
Br^-	0.92		1.1	0.91			0.95		1.2	1.00		1.6
I^-	0.86		0.7	0.88		0.8	0.94		0.8	1.00		1.2
O^{2-}	0.97	0.96	1.6	1.04	1.028	1.7	1.08	1.082	1.7	1.17	1.150	2.1
S^{2-}	0.94	0.71 [a]	0.7		0.73 [a]	0.8	0.964	0.95 [c] 0.89 [b] 0.80 [a]	0.8	1.015	1.022	1.2
Se^{2-}		0.71 [a]	0.6		0.69 [a]	0.7		0.77 [a]	0.7	0.982	1.00	1.1
Te^{2-}										0.75		0.9
OH^-	0.95			0.97			1.02			1.10		

	Zn and Mg			Cd and Mg			Cd and Ca			In and Sc		
	$\bar{R}_v$	$\bar{R}_d$	$\Delta\varkappa$	$\bar{R}_v$	$\bar{R}_d$	$\Delta\varkappa$	$\bar{R}_v$	$\bar{R}_d$	$\Delta\varkappa$	$\bar{R}_v$	$\bar{R}_d$	$\Delta\varkappa$
F^-	1.077	1.061	2.4	1.29	1.26		0.96	0.95	2.3	1.10		2.2
Cl^-	1.08		1.6	1.10			0.92		1.5			
Br^-												
I^-	1.00		1.0	1.04			0.88		0.9			
O^{2-}	1.05	1.034	1.9	1.39	1.33		0.94		1.8	1.08		1.7
S^{2-}				1.023							1.04	0.8
Se^{2-}												
Te^{2-}	0.85		0.7									
OH^-	0.99			1.2			0.91			1.04		

a) NiAs structure.
b) Distorted NiAs.
c) Olivine structure.

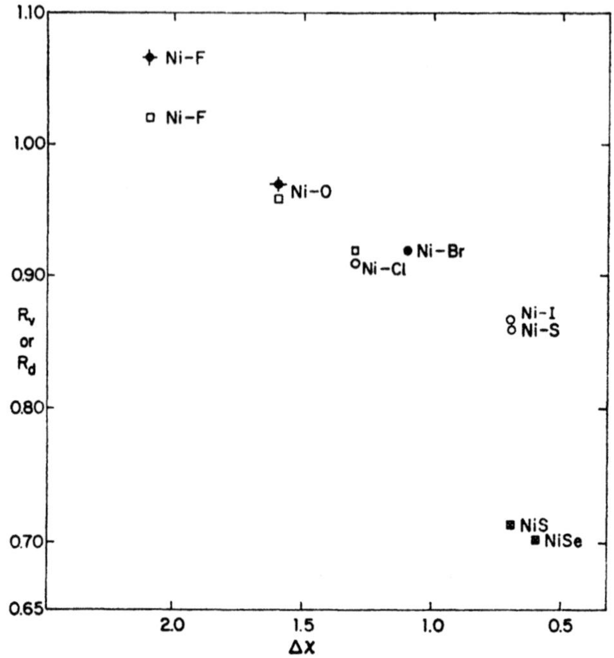

Fig. 1. R_v *vs.* $\Delta\varkappa$ for Ni compounds

● — R_v derived from many isotypic Ni and Mg compounds

○ — R_v derived from only a few isotypic Ni and Mg compounds

□ — R_d derived from several isotypic compounds

⊠ — R_d derived from compounds with similar M coordination but different structures

for Mn; and S and Te for Zn. The slopes of the R_v—$\Delta\varkappa$ curves are similar for Ni, Fe, Mn, Co, and Cr. For Zn, Cd, and In the slopes are significantly less. This difference in slope may be due to the difference in covalence of hybrid orbitals formed from metal *d* orbitals *vs.* metal *s*—*p* orbitals.

The relationship between R_v and $\Delta\varkappa$ is approximately linear if one neglects the NiAs phases containing S, Se and Te. Unfortunately, the sulfides, selenides and tellurides of Ni, Fe, Mn and Co are seldom isotypic with those of Mg so that true values of R_v are not available for many of the highly covalent selenides and tellurides. It is apparent that R_v is structure dependent for highly covalent compounds *e.g.*, R_v for Fe^{2+}—S is 0.95 (M_2GeS_4 olivines), 0.89 (distorted NiAs structure) and 0.80 (NiAs structure). However, the two examples of isotypic tellurides MnTe/MgTe

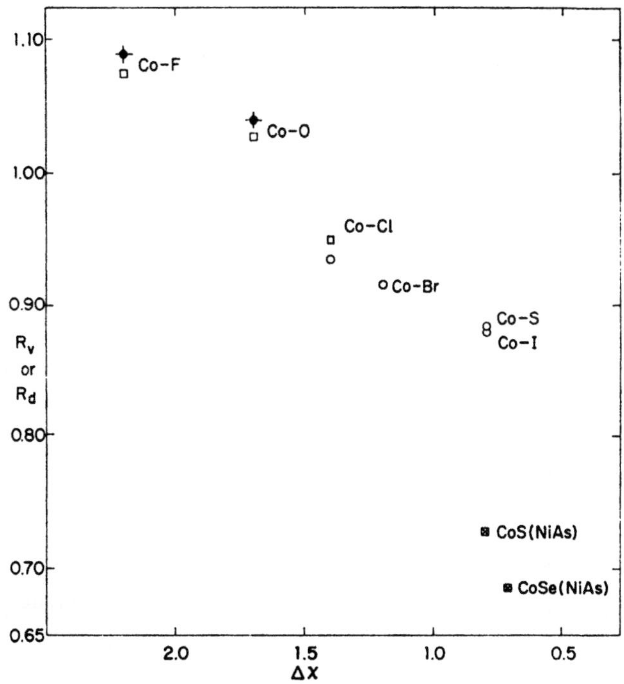

Fig. 2. R_v *vs.* $\Delta\varkappa$ for Co compounds

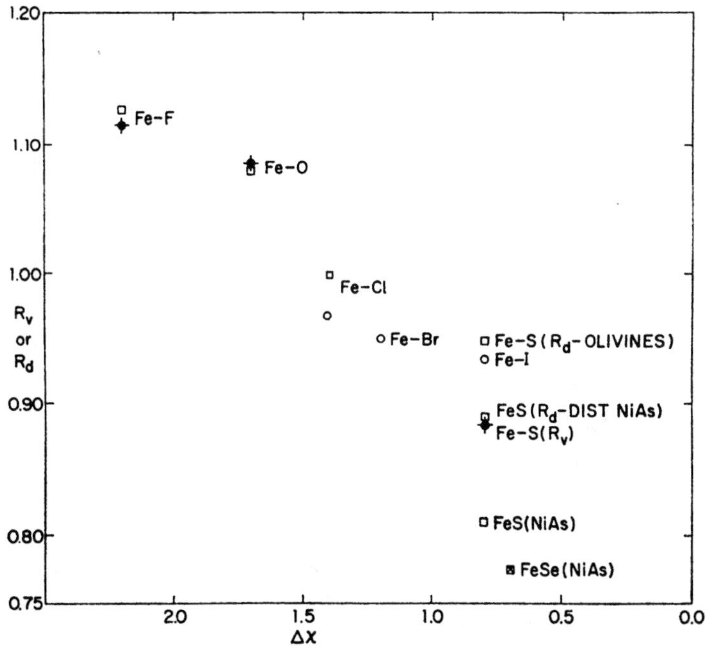

Fig. 3. R_v *vs.* $\Delta\varkappa$ for Fe compounds

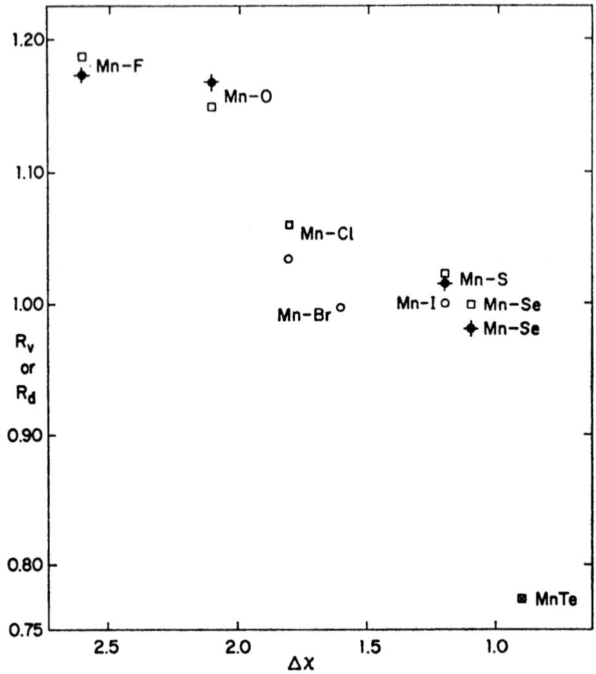

Fig. 4. R_v *vs.* $\Delta\varkappa$ for Mn compounds

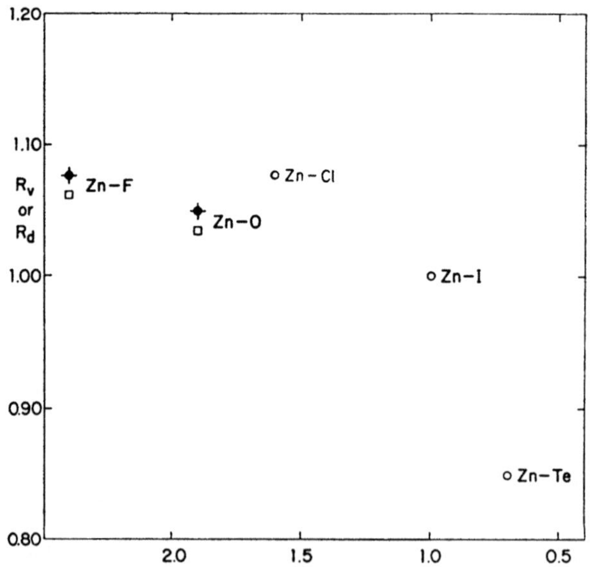

Fig. 5. R_v *vs.* $\Delta\varkappa$ for Zn compounds

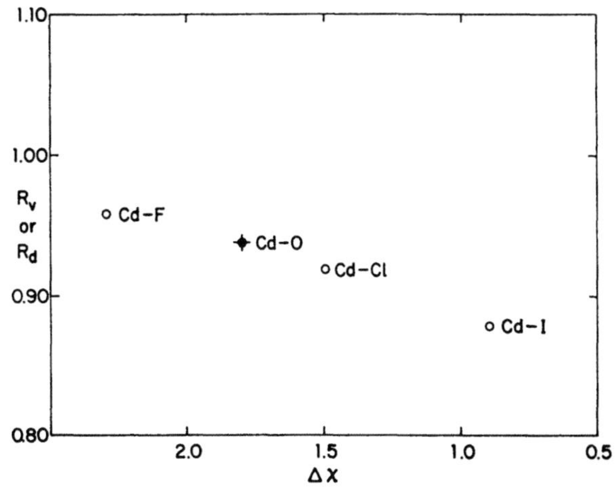

Fig. 6. R_v *vs.* $\Delta\varkappa$ for Cd compounds

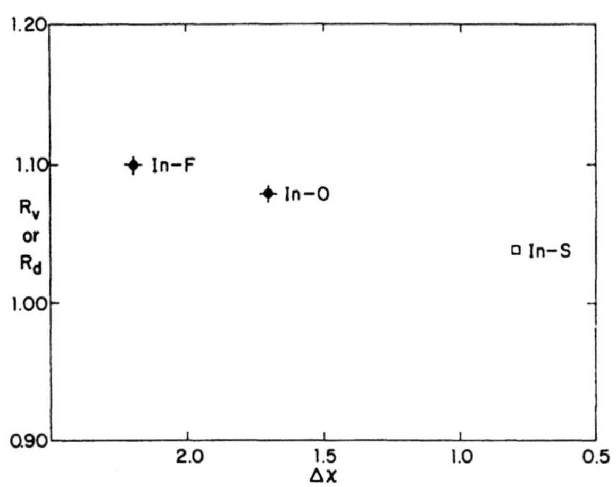

Fig. 7. R_v *vs.* $\Delta\varkappa$ for In compounds

and ZnTe/MgTe lie far below the line extrapolated for Mn—X and Zn—X compounds. This suggests a parabolic R_v *vs* $\Delta\varkappa$ plot.

Covalency effects on cell dimension *vs* ionic radii plots manifest themselves in a lowering of the line joining the more covalent ions relative to a line joining the Mg and Ca compounds. Thus, in fluorides, Ni

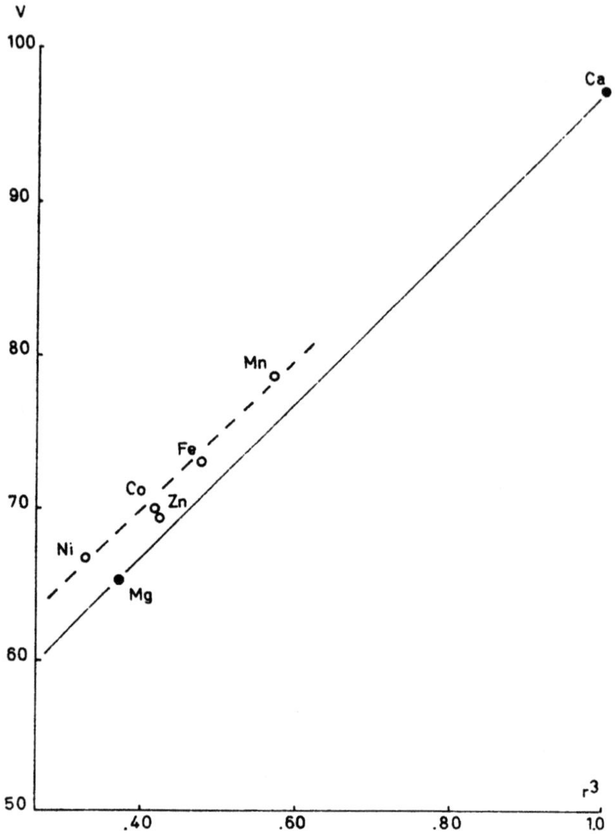

Fig. 8. Unit cell volume *vs.* r^3 for MF_2 rutile compounds

compounds have larger V than corresponding Mg compounds (Fig. 8), whereas in oxides and sulfides (Figs. 9 and 10) the Mg compounds have a larger V. Similarly, in fluorides and oxides, Mn compounds have a much greater V than the corresponding Mg compounds (Figs. 8 and 9), whereas in sulfides the Mg and Mn compounds have approximately equal V (Fig. 10). In selenides, the volumes of the Mn compounds are smaller than those of Mg.

The same effect can be seen in ternary oxides $M_xY_yO_z$, *e.g.*, $M_3V_2O_8$ (Fig. 11), MV_2O_6 (*Gondrand et al.* 1974), $MMoO_4$ (Fig. 12) and MWO_4 (Fig. 13). In each case the line connecting V of the compounds containing the more electronegative elements Ni, Co, Zn, Mn, and Cd is lower than

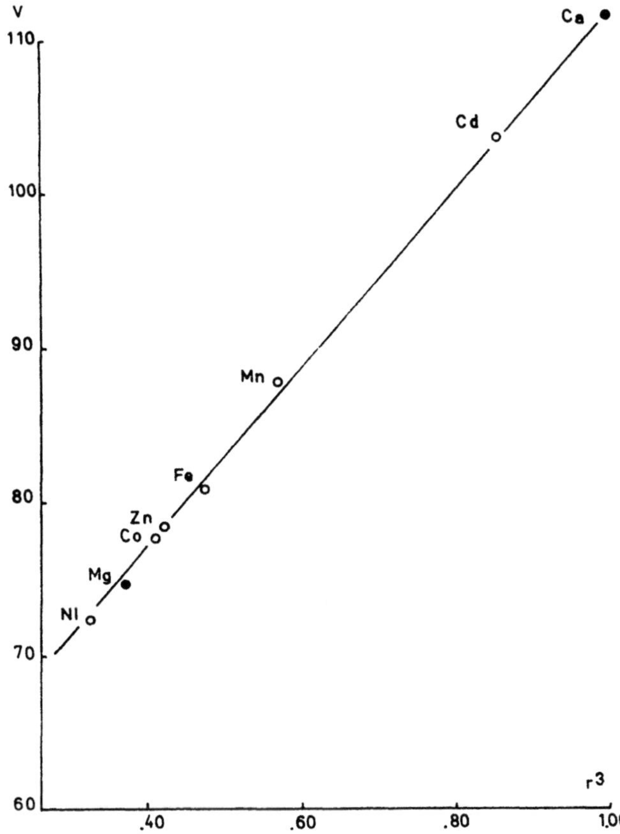

Fig. 9. Unit cell volume *vs.* r^3 for MO rocksalt compounds

the line connecting compounds containing the less electronegative Mg and Ca. Note that the radii used in these plots are the radii which give a linear plot for the simple oxides in Fig. 9. We attribute the reduced volume of the $M_xY_yO_z$ compounds to the increased covalence of the M—O bond when the O is also bonded to more electronegative cations such as W^{6+}, Mo^{6+}, or V^{5+}. There is apparently an analogy between the anions S^{2-}, Se^{2-} and the anion groups WO_4^{2-}, MoO_4^{2-}, VO_4^{3-}, VO_5^{5-} and VO_6^{7-}. The position of the Mg compounds *vs* those of Ni and Fe indicates that the effective electronegativity of oxygen in these anion groups is intermediate between that in simple oxides and sulfides. This concept helps to explain the different degree of covalence found for Mn—O in MnO and $MnCO_3$, to be discussed later.

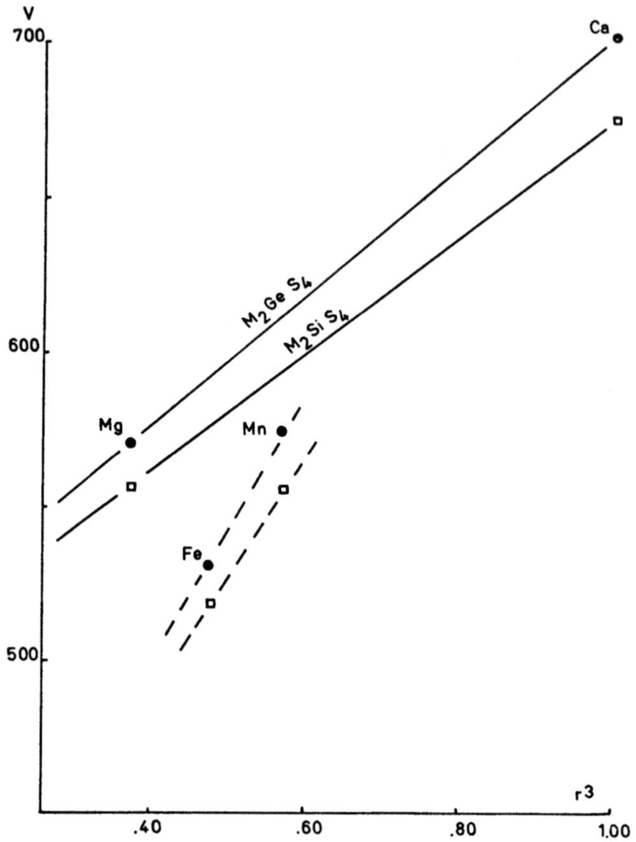

Fig. 10. Unit cell volume *vs.* r_M^3 for M_2SiS_4 and M_2GeS_4 olivine compounds

However, the concept of covalence as defined by electronegativity is not sufficient to explain quantitatively the variation in cell volumes of $M_xY_yO_z$ compounds. Table 4 shows that the degree of covalent shortening is greater for vanadates, than molybdates and tungstates, in contradiction to electronegativity values. This same difference was found in an analysis of tetrahedral VO_4^{3-} and MoO_4^{3-} distances by *Shannon* (1971). It appears that the degree of covalent shortening is somehow related to the degree of double bond formation if we use the range of distances in a VO_6 or MoO_6 octahedron as a measure of this. Thus, we see Mo^{6+}—O distances varying from 1.69 → 2.4 for a typical octahedral distance of 1.96 Å. Octahedral V^{5+}—O distances vary from 1.58 → 2.5 Å for a typical V^{5+}—O octahedral distance of 1.90 Å. The greater range of distances

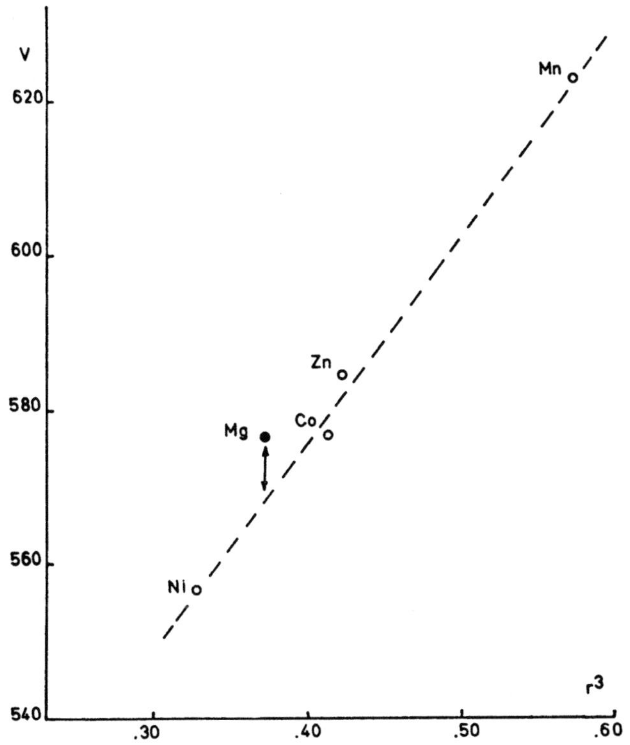

Fig. 11. Unit cell volume *vs.* r_M^3 for $M_3V_2O_8$ compounds

Table 4. Reduction of cell volumes of ternary transition metal oxides ($M_xY_yO_z$) relative to the volume of ternary Mg-containing oxides

Composition	Structure type	Deviation	$\varkappa_y$	Reference
MWO_4	Scheelite	0.9%	2.2	[a]
$MMoO_4$	Scheelite	1.1%	2.3	[a]
$M_3V_2O_8$	$Mg_3V_2O_8$	1.5%	1.9	[b]
MV_2O_6	Columbite	1.7%	1.9	[c]
MV_2O_6	Brannerite	2.3%	1.9	[c]

[a] *Sleight, A. W.:* Acta Cryst. *B 28*, 2899 (1972).
[b] *Shannon, R. D.:* unpublished data.
[c] *Gondrand, M., Collomb, A., Joubert, J. C., Shannon, R. D.:* J. Solid State Chem. 11, to be published (1974).

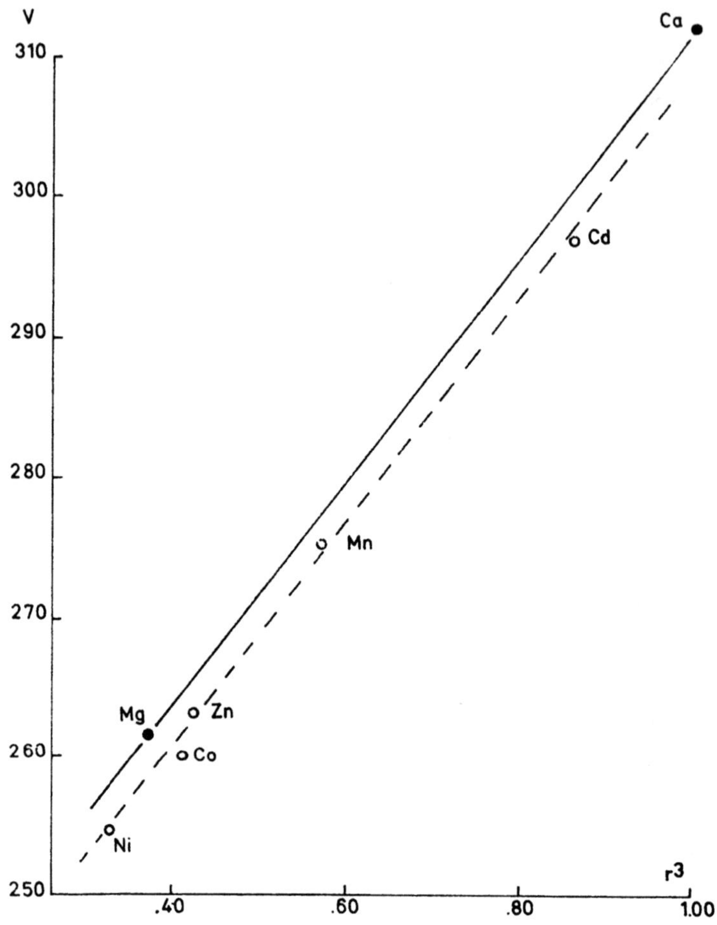

Fig. 12. Unit cell volume *vs.* r_M^3 for $MMoO_4$ scheelite compounds

for the V^{5+}—O bonds suggests a greater degree of double bond formation than in the Mo^{6+}—O group.

We expect greater covalent shortening in $M_xY_yO_z$ series when Y is tetrahedral and when Y is present in greater concentration. The results in Table 4 are in agreement with these considerations. Thus, we see greater shortening for MV_2O_6 brannerite than for MV_2O_6 columbite (decreased V^{5+} coordination in brannerite). In the $M_3V_2O_8$ series the effects of the tetrahedral coordination compensates for the decrease in V content. Finally, the larger degree of shortening in molybdates over

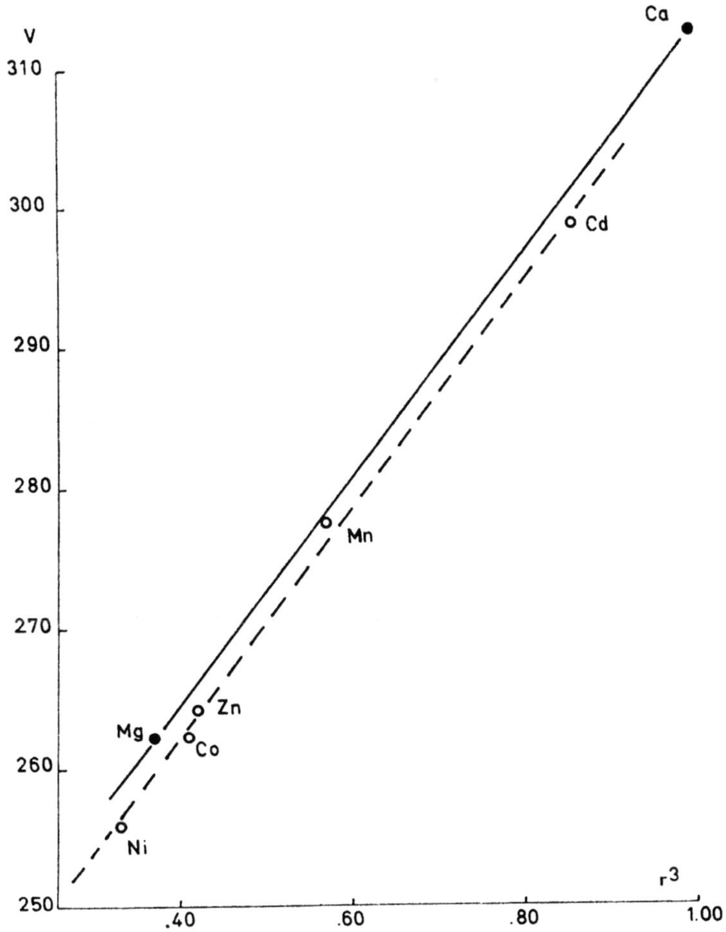

Fig. 13. Unit cell volume *vs.* r_M^3 for MWO_4 scheelite compounds

tungstates is in agreement with Sleight's conclusion (*Sleight*, 1972) that the MoO_4^{2-} group is slightly more covalent than the WO_4^{2-} group.

Biggar (1969) calculated unit cell volume ratios for isotypic Ni and Mg compounds in order to revise the Mg^{2+} and Ni^{2+} ionic radii of *Ahrens* (1952) and *Pauling* (1960). He concluded that $r(Ni^{2+}) = 0.97 \times r(Mg^{2+})$ in oxides and halides. It is interesting to note that *Biggar* found ratios for the halides F, Cl, Br, and I in agreement with the electronegativity dependence in this paper ($\bar{R}_V = 1.060, 0.917, 0.919$, and 0.836 respectively) but he assumed that the deviations from 0.970 were the result of faulty data. We believe that these deviations are real and are caused by different degrees of covalence.

The previous literature on the effects of partial covalence on interatomic distances is contradictory. *Pauling* (1960) cites the examples of CuF, BeO, AlN, and SiC where observed bond lengths are shorter than the sum of the covalent radii. He attributes these differences to partial ionic character and thus implies that partial ionic character shortens covalent bonds. This conclusion is in accord with the *Schoemaker—Stevenson* (1941) rule $D_{A-B} = r_A + r_B - C\,|\varkappa_A - \varkappa_B|$ where D = interatomic distance between A and B, r_A and r_B = covalent radii of A and B, $\varkappa_A$ and $\varkappa_B$ = electronegativity of A and B and C = constant.

On the other hand *Wells* (1949) provided strong evidence against the validity of the *Schoemaker—Stevenson* rule. Futhermore, if for the pairs $^{VI}Ni^{2+}$—X, $^{VI}Fe^{2+}$—X and $^{VI}Co^{2+}$—X the sums of *Pauling's* crystal and covalent radii are compared (Table 5), it is clear that the covalent distances are predicted to be shorter than the ionic distances. *Roth* (1967) in a comparison of the sums of ionic radii with observed

Table 5. Comparison of sums of covalent and ionic radii

Bond	Covalent sum[a]	Ionic sum[a]	Ionic sum[b]	Average observed distance[c]
Ni—F	2.03	2.08	2.02	2.00
Cl	2.38	2.53	2.50	
Br	2.50	2.67	2.65	
I	2.67	2.88	2.89	
O	2.05	2.12	2.09	2.084
S	2.43	2.56	2.53	
Se	2.53	2.70	2.67	
Fe—F	1.87	2.12	2.07	2.08
Cl	2.22	2.57	2.55	
Br	2.34	2.71	2.70	
I	2.51	2.92	2.94	
O	1.89	2.16	2.14	2.16
S	2.27	2.60	2.58	2.54
Se	2.37	2.74	2.72	
Co—F	1.96	2.10	2.05	2.04
Cl	2.31	2.55	2.53	
Br	2.43	2.69	2.68	
I	2.60	2.90	2.92	
O	1.98	2.14	2.10	2.133
S	2.36	2.58	2.56	
Se	2.46	2.72	2.70	

a) *Pauling* (1960).
b) *Ahrens* (1952).
c) Fluoride distances were taken from rutile compounds; oxide distances were taken from rocksalt compounds.

distances in the chalcogenides of Zn, Cd, and Hg also concluded that interatomic distances decrease as the amount of covalency increases. Similarly, *Pauling* also states that observed distances in FeI_2 (2.88 Å) are greater than the sum of the covalent radii (2.58 Å). The results for FeI_2 and those found by *Roth* and in Table 5 correspond to our conclusions found by comparing ratios of distances and volumes as in Table 3.

Further evidence for covalency effects comes from a comparison of interatomic distances and Y—O symmetric stretching frequencies (Table 6) in the MYO_4 scheelite compounds where Y = Mo or W. As the covalent character of the M—O bond increases (as measured by $\varkappa_M$ in Table 6) and thus that of the Y—O bond decreases, the mean Y—O distance increases. This increase in Y—O distance is accompanied by a decrease in the symmetric stretching frequency of the YO_4^- group. A similar relationship between M—O covalency and IO_4^- stretching frequencies exists for the MIO_4 scheelite compounds (*Tarte*, 1973).

Table 6. Interatomic distances in MXO_4 scheelite compounds

Compound	$\varkappa_M$	Mean Mo^{6+}—O or W^{6+}—O distance, Å	ν_1[e]), cm^{-1}
$SrMoO_4$	0.95	1.766 ± 0.005[a])	887
$CaMoO_4$	1.00	1.757 ± 0.005[a])	879
$PbMoO_4$	1.87	1.772 ± 0.006[b])	869
$BaWO_4$	0.89	1.781 ± 0.003[a])	922
$SrWO_4$	0.95	1.779 ± 0.003[a])	919
$CaWO_4$	1.00	1.786 ± 0.003[c])	910
$PbWO_4$	1.87	1.804[d])	902

a) *Gurmen, E., Daniels, E., King, J. S.:* J. Chem. Phys. *55*, 1093 (1971).
b) *Leciejewicz, J.:* Z. Krist. *121*, 158 (1965).
c) *Zalkin, A., Templeton, D. H.:* J. Chem. Phys. *40*, 501 (1964).
d) *Plakhov, G. F., Pobedimskaya, E., Simonov, M., Belov, N. V.:* Sov. Phys. Cryst. *15*, 928 (1971).
e) *Liegeois—Duyckaerts, M., Tarte, P.:* Spectrochim Acta, *28A*, 2037 (1972).

III. Correlations between R_v, Magnetic Moment Reduction and Mössbauer Isomer Shift

A. Introduction

In an ionic compound, the partial covalence of a bond formed between a transition metal ion and its ligand modifies the magnetic properties of the cation. It can be seen, for example, that if electrons were transferred from the ligands to partially occupied $3d$ orbitals of the central ion, the magnetic properties of the central ion which depend on the $3d$ electrons are modified. *Owen* and *Thornley* (1966) discussed in detail the effects of covalence on crystal field splitting, orbital magnetic moment, ligand hyperfine structure, charge transfer, neutron diffraction and exchange interactions. They cite the following evidence for the effects of covalence:

1. the increase in crystal field splitting, Δ, between the t_{2g} and e_g levels (*Anderson*, 1963)
2. the reduction of the orbital magnetic moment of the central ion and the spin-orbit coupling of the ligand (*Stevens*, 1953; *Misetich* and *Buch*, 1964) and
3. the reduction of the Raccah parameters.

Electron spin resonance, nuclear magnetic resonance, and neutron diffraction methods allow a quantitative determination of the degree of covalence. The reasonance methods utilize the hyperfine interaction between the spin of the transferred electrons and the nuclear spin of the ligands (*Stevens*, 1953), whereas the neutron diffraction methods use the reduction of spin of the metallic ion as well as the expansion of the form factor (*Hubbard* and *Marshall*, 1965). The Mössbauer isomer shift which depends on the total electron density of the nucleus (*Walker et al.*, 1961; *Danon*, 1966) can be used in the case of Fe. It will be particularly influenced by transfer to the empty $4s$ orbitals, but transfer to $3d$ orbitals will indirectly influence the $1s$, $2s$, and $3s$ electron density at the nucleus.

Given the apparent relationship between covalence and contraction of the unit cell volume described previously, it should be possible to relate R_V to the reduction in magnetic moment found by resonance and neutron diffraction. In this we are limited to the cations Mn^{2+}, Fe^{2+}, Co^{2+}, and Ni^{2+} in octahedral coordination.

B. Theory

One can visualize the effects of covalence on magnetic properties using a simple molecular orbital scheme. In the usual notation the orbitals considered for the transition metal M and the ligand X are:

		$M(3d^n)$		
empty orbitals	$4p_x$	$4p_y$	$4p_z$	(σ or π)
		$4s$		(σ)
partially filled orbitals	$3d_{z^2}$	$3d_{x^2-y^2}$		(σ) e_g
	$3d_{xy}$	$3d_{yz}$	$3d_{xz}$	(π) t_{2g}

		$X(2s^2, 2p^6)$		
filled orbitals	$\{2p_x$	$2p_y$	$2p_z$	(σ or π)
		$2s$		(σ)

A partially covalent bond between M and X can be formed only by orbitals for which overlap is possible. For example, if the z axis points along the line joining M and X, the metal $3d_{z^2}$ orbitals and ligand $2p_z$ or $2s$ can form a σ bond, and $3d_{yz}$ and $2p_y$ a π bond. However, $3d_{yz}$ and $2p_x$ or $2s$ with zero overlap cannot form a bond. The p or s orbitals of the ligand contain 2 electrons ($\uparrow\downarrow$); certain $3d$ metal orbitals contain a single electron ($\uparrow$). Covalency consists of transferring part of a p or s electron of spin ($\downarrow$) onto a $3d$ orbital of the metal. A coefficient f is then defined according to the bond type and corresponds to the percentage of spin transferred:

$$f_\sigma, f_\pi, f_s \text{ or } A_\sigma^2, A_\pi^2, \text{ or } A_s^2$$

Of course, electrons can also be transferred into empty $3d$ or $4s$ orbitals; *Rimmer* (1964) noted the importance of this contribution in certain cases.

C. Results

For Mn^{2+}, Co^{2+}, and Ni^{2+} compounds we have neglected electron transfer to empty cation orbitals because it is difficult to determine this quantity. We are thus concerned only with covalence involving unpaired electrons. We have plotted R_V or R_d defined in the first section *vs* the percentage of total spin transferred as calculated from spin transfer coefficients where they were known, or taken from values of spin reduction according to magnetic structures determined by neutron diffraction. Unfortunately, in this latter case the presence of spin-orbit coupling in many compounds containing Co^{2+} or Ni^{2+} reduces the number of possible examples. Furthermore, very few authors have determined the spin reduction. Consequently, the results obtained indirectly from magnetic structures are less reliable than those from spin transfer coefficient work where covalency determination was the major goal.

1. Mn^{2+}

Mn^{2+} has the electronic configuration $3d^5$ ($t_{2g}^3 e_g^2$). The total transfer of spin, ΔS, is given by the relation:

$$\Delta S = \frac{6}{5}(f_\sigma + 2f_\pi + f_s) \cdot S_0$$

where S_0 is the total spin; the coefficient 6/5 refers to the fact that transfer takes place from 6 ligands to 5 unpaired electron orbitals; and the factor $2f_\pi$ appears because each $3d_\pi$ orbital points towards 2 ligand p_π orbitals. Table 7 and Fig. 14a show the plot of R_V and R_d *vs.* the relative reduction of spin observed, $\Delta S/S_0$.

Table 7. Spin reduction of Mn^{2+} compounds

Compound	R_V	R_d	$f_\sigma\%$	$f_\pi\%$	$f_s\%$	$\Delta S/S_0\%$	References
MnF_2	1.20	1.20	1.2	0.8	0.5	3.9	*Nathans et al.* (1964) *Rimmer* (1964)
MnO	1.18	1.17			0.7	4.2	*O'Reilly et al.* (1964) *Fender et al.* (1968)
$MnCO_3$	1.12					8.0	*Lindgård et al.* (1969)
$TlMnCl_3$	1.06					7.0 (± 1.0)	*Malamud et al.* (1970)
α MnS	1.01	1.01				9.1	*Fender et al.* (1968)
MnI_2	1.00					8.0 (± 1.0)	*Cable et al.* (1962)

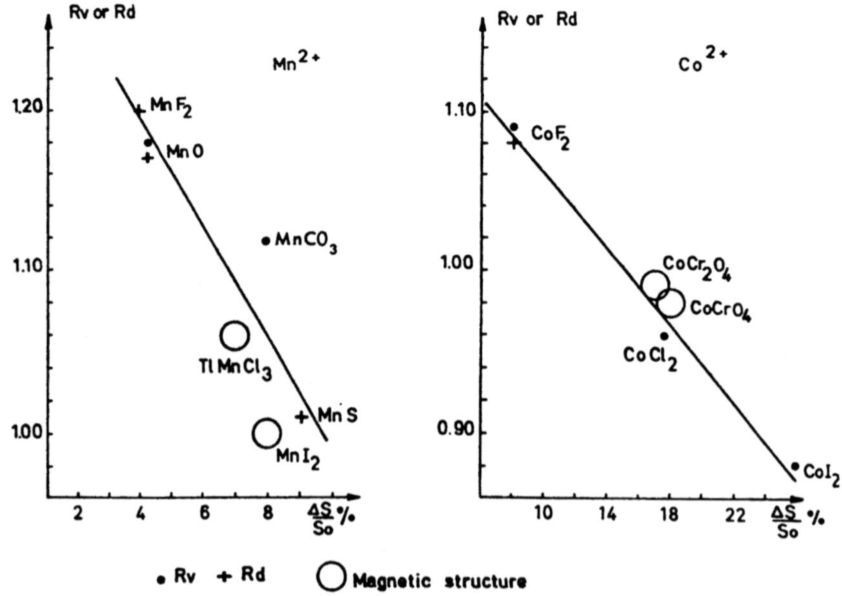

Fig. 14a. R_V *vs.* percent spin reduction in Mn^{2+} compounds

Fig. 14b. R_V *vs.* percent spin reduction in Co^{2+} compounds

2. Co^{2+}

Co^{2+} has the configuration $3d^7(t_{2g}^5 e_g^2)$. The total spin transfer ΔS is given by:

$$\Delta S = \frac{6}{3}\left(f_\sigma + \frac{2}{3} f_\pi + f_s\right) \cdot S_o$$

Here $2f_\pi$ must be multiplied by 1/3 because there is only 1 unpaired electron in the t_{2g} orbitals instead of 3 as in Mn^{2+}. Table 8 and Fig. 14b show the relationship between R_v and total spin transfer.

Table 8. Spin reduction of Co^{2+} compounds

Compound	R_v	R_d	f_σ%	f_π%	f_s%	$\Delta S/S_o$%	References
CoF_2	1.09	1.08	2.4	2.4	0.5	9.1	*Thornley* (1962)
$CoCr_2O_4$	0.99					17.0 (± 2.0)	*Plumier* (1968)
$CoCrO_4$	0.98					18.0 (± 2.0)	*Pernet et al.* (1969)
$CoCl_2$	0.96		5.0	5.0	0.6	17.6	*Thornley* (1962) *Fender et al.* (1967)
$CoBr_2$			5.3	5.3	0.5	18.8	*Windsor et al.* (1962)
CoI_2	0.88		7.5	7.5	0.5	26.0	*Windsor et al.* (1962)

3. Ni^{2+}

Ni^{2+} has the electron configuration $3d^8(t_{2g}^6 e_g^2)$; there are thus two unpaired e_g electrons and $f_\pi = 0$

$$\Delta S = \frac{6}{2}\,(f_\sigma + f_s) \cdot S_o$$

Table 9 and Fig. 14c, d show the relationship between R_v and ΔS.

Table 9. Spin reduction of Ni^{2+} compounds

Compound	R_v	R_d	f_σ%	f_s%	$\Delta S/S_o$%	References
$KNiF_3$	1.03	1.03	3.8	0.5	12.9	*Nathans et al.* (1964) *Rimmer* (1964)
NiO	0.97	0.97			18.0	*Alperin et al.* (1961)
$RbNiCl_3$	0.95				25.0 (± 3.0)	*Minkiewicz* (1970)
Cr_2NiS_4	0.92				30.0 (± 3.0)	*Andron et al.* (1966)

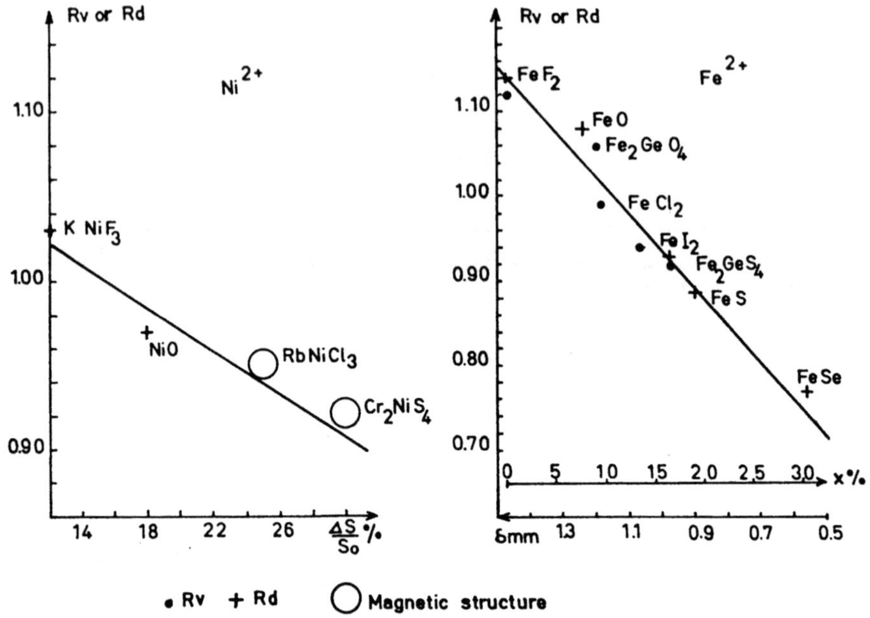

Fig. 14c. R_V *vs.* percent spin reduction in Ni^{2+} compounds

Fig. 14d. R_V *vs.* Mössbauer isomer shift in Fe^{2+} compounds

4. Fe^{2+}

High spin Fe^{2+} has the configuration $3d^6(t_{2g}^4 e_g^2)$. Although we could examine the relationship between R_V and ΔS as for Mn^{2+}, Co^{2+}, and Ni^{2+}, we prefer in this case to look at R_V as a function of the occupation of the combination of $3d$ and $4s$ orbital by ligand electrons which is measured by the Mössbauer isomer shift. In general, the coefficient f_σ, f_π, and f_s are not known for Fe^{2+}. In addition, possible spin-orbit coupling makes it difficult to determine the spin reduction by magnetic structures. However, the isomer shift allows us to determine approximately the occupancy of the $4s$ orbitals and there are many experimental results available.

The isomer shift δ is related to the density of s electrons at the nucleus of the source S and the absorber A by the expression:

$$\delta = \alpha \cdot (|\Phi(O)_A{}^2 - |\Phi(O)_S|^2)$$

where α is a constant of proportionality equal to:

$$\alpha = \frac{4\pi}{5} 2e^2 R^2 S'(Z) \frac{\Delta R}{R}$$

with:

Z = number of electrons

R = radius of the nucleus

ΔR = relative variation of R between the excited and ground state

$S'(Z)$ = relativistic correction factor

and $|\Phi(O)_A|^2$ and $|\Phi(O)_S|^2$ are the two s electron densities at the nucleus of the absorber and source, respectively. If there were only electron transfer from ligands to the $4s$ orbitals of Fe^{2+} by covalence, *i.e.*, when the electron configuration can be written $3d^6 4s^x$, there is a change of electron density s at the nucleus. If this compound is the absorber:

$$|\Phi(O)_A|^2 = \sum_{n=1}^{3} |\Phi_{ns}(O)|^2 + x\,|\Phi_{4s}(O)|^2$$

and δ varies linearly with x. *Walker et al.* (1961) have proposed a scale of x as a function of isomer shift δ for Fe^{+2}.

Table 10 and Fig. 14d show the relationship between R_v and δ where we have used the x proposed by *Walker* as abcissa.

Table 10. Mössbauer isomer shifts in Fe^{2+} compounds

Compound	R_v	R_d	$\delta\,mm/s$ [a]	x% ($4s^x$)	References
FeF_2	1.12	1.14	1.47	0	*Wertheim et al.* (1967)
FeO	1.08	1.08	1.24	7.7	*Romanov* (1972)
Fe_2GeO_4	1.06		1.20	9.0	*Imbert* (1966)
$FeCl_2$	0.99		1.19	9.3	*Ôno et al.* (1964)
FeI_2	0.94		1.07	13.3	*Hazony et al.* (1968)
Fe_2GeS_4	0.92	0.93	0.98	16.3	*Meyer* (1974)
FeS		0.89	0.90	19.0	*Ôno et al.* (1962)
FeSe		0.77	0.56	30.3	*Ôno et al.* (1962)

[a] Isomer shifts δ are relative to stainless steel. The shifts contain contributions from the second order *Doppler* shift and the zero point energy, in addition to the isomer shift discussed here, as in the *Walker* publication.

D. Discussion

We can now make the following remarks concerning bonds between the transition metals Mn^{2+}, Co^{2+}, Ni^{2+}, and Fe^{2+} and ligands such as F, Cl, Br, O, S, and Se.

1. Qualitatively, the ratio R_v or R_d decreases as the covalence increases, *i.e.*, transfer of electrons from ligands to the cation reduces the

cation-anion distance and contracts the cell. This result justifies our previous remark concerning Fe_2GeS_4 (*Vincent et al.*, 1973) where we have attributed the apparent cell contraction and the spin reduction to the Fe—S bond covalence.

2. More quantitatively, it appears to a first approximation that the unit cell contraction of a compound containing Mn^{2+}, Co^{2+}, Ni^{2+}, or Fe^{2+}, relative to the isomorphous Mg^{+2} compound, is a linear function of the $\Delta\varkappa$ of the metal-ligand bond if we neglect selenides and tellurides. Inclusion of these more covalent compounds indicates a greater dependence on $\Delta\varkappa$.

3. Figs. 2 and 4 show that the cell contraction is almost the same for Mn^{2+} and Co^{2+} as covalency increases from fluorides to iodides or sulfides (thus the total $3d+4s$ covalence is identical for both). However, the covalence of the $3d$ orbitals is 2 to 3 times smaller for Mn^{2+} (see Figs. 14a and 14b). Therefore Mn^{2+}—X covalence results mostly from spin transfer to empty $4s$ orbitals, unlike Co^{2+}. This agrees with the conclusion of *Rimmer* (1964).

It is interesting to note that the large difference in covalency parameters for the Mn^{2+}—O bond in MnO (*Nathans et al.*, 1964) *vs* the Mn^{2+}—O bond in $MnCO_3$ (*Lindgard* and *Marshall*, 1969) agrees well with the spin transfer anticipated from R_v and R_d. This covalency difference for an M—O bond in ternary oxides $M_xY_yO_z$, which is dependent on the nature of the Y—O bonds [in this case (Mn—O)—Mn for MnO and (Mn—O)-C for $MnCO_3$], was also shown to exist for other oxides (see Table 4).

IV. Conclusions

1. The covalency contraction parameter, R_v, which measures the volume of a transition metal compound M_mX_n relative to the volume of Mg_mX_n, is proportional to the electronegativity of X and thus decreases as the covalence of the M—X bond increases.

2. This concept explains the relative differences in cell volume for certain isotypic pairs such as NiF_2—MgF_2 and NiO—MgO, where the volume of the Ni compound is larger than that of Mg for fluorides and smaller for oxides. Similarly, for the pairs FeF_2—MgF_2 and Fe_2GeS_4—Mg_2GeS_4, it explains the fact that FeF_2 has a larger cell than MgF_2, whereas Mg_2GeS_4 has a larger cell than Fe_2GeS_4.

3. In ternary oxides, $M_xY_yO_z$, the unit cell volume of compounds in which M = transition metal *vs* the volume of $Mg_xY_yO_z$ is a function of the Y—O_n group, being smaller as: the ratio of $x : y$ decreases; the coordination of Y decreases; and the Y—O covalence increases. Thus, the

ratio R_V is smaller for: CoV_2O_6 *vs* $Co_3V_2O_8$, NiV_2O_6 (brannerite) *vs* NiV_2O_6 (columbite) and $MnCO_3$ *vs* MnO.

4. MXO_4 scheelite compounds show a correlation between mean X—O distance, covalent character of the X—O bond, and symmetric stretching frequency of the XO_4 group.

5. There is an approximately linear relationship between R_V and spin transfer coefficients determined from electron and nuclear magnetic resonance and neutron diffraction, *i.e.*, a contraction of the unit cell accompanies the transfer of spin from transition metal to the ligands.

6. An approximately linear relationship also exists between R_V and the Fe^{2+} Mössbauer isomer shift, *i.e.*, between shortening of Fe^{2+}—X bonds and increased electron transfer to the 4*s* orbitals of Fe^{2+}.

7. The differences in spin transfer observed in MnO and $MnCO_3$ are consistent with the respective values of R_V for the two compounds.

Appendix

If one assumes to a first approximation that the contraction of the cell is a linear function of the covalence of the metal-ligand bond, one can derive for the bivalent ions the following equations:

Mn^{2+}

$$\frac{\Delta S}{S}(\%) = -30.0R + 39.7$$

Co^{2+}

$$\frac{\Delta S}{S}(\%) = -83.6R + 98.8$$

Ni^{2+}

$$\frac{\Delta S}{S}(\%) = -157R + 173$$

Fe^{2+}

$$\delta(mm/SS) = 2.26R - 1.12$$

i.e., $x(\%) = -75.7R + 86.6$
according to *Walker*

where $R = R_V$ or R_d.

Acknowledgments. We would like to thank Drs. *P. Tarte, V. Johnson, J. F. Weiher, R. J. Bouchard, Y. Gros, J. M. D. Coey* and *E. F. Bertaut* for helpful discussions and for critically reviewing the manuscript prior to publication.

References

Ahrens, L. H.: Geochim. Acta *2*, 155 (1952).
Allred, A. L.: J. Inorg. Nucl. Chem. *17*, 215 (1961).
Alperin, H. A.: J. Phys. Soc. Japan, Suppl. *17*, 12 (1962).
Anderson, P. W.: Solid State Physics (Ed. F. Seitz and D. Turnbull) *14*, 99. New. York: Academic Press 1963.
Andron, B., Bertaut, E. F.: J. Phys. *27*, 619 (1966).
Banks, E., Greenblatt, M., Post, B.: Inorg. Chem. *9*, 2259 (1970).
Batsonov, S.: Russ. Chem. Rev. *37*, 332 (1968).
Biggar, G. M.: Mineral. Mag. *37*, 299—300 (1969).
Cable, J. W., Wilkinson, M. K., Wollan, E. O., Koehler, W. C.: Phys. Rev. *125*, 1860 (1962).
Danon, J.: Tech. Rept. Ser. Intern. At. Energy Agency *50*, 89 (1966).
Donnay, J. D. H., Ondik, H.: (1973) Crystal Data, Vol. II Inorganic Compounds, Nat'l Bureau of Stds (1973).
Fender, B. E., Jacobson, A. J., Wedgwood, F. A.: J. Chem. Phys. *48*, 990 (1968).
Gondrand, M., Collomb, A., Joubert, J. C., Shannon, R. D.: J. Solid St. Chem. *11*, to be published (1974).
Gordy, W., Thomas, W. J.: J. Chem. Phys. *24*, 439 (1956).
Hazony, Y., Axtmann, R. C., Hurley, J. W.: Chem. Phys. Letters *2*, 673 (1968).
Horita, H., Hirahara, E.: J. Phys. Soc., Japan *21*, 1447 (1966).
Hubbard, J., Marshall, W.: Proc. Phys. Soc. *86*, 561 (1965).
Imbert, P.: Compt. Rend. *263B*, 184 (1966).
Lindgard, P. A., Marshall, W.: J. Phys. C. (Solid State Physics) *2*, 276 (1969).
Longo, J. M., Kafalas, J. A.: J. Solid St. Chem. *1*, 103—108 (1969).
Malamud, M., Pinto, H., Schacher, G., Makovsky, J., Shaked, H.: Tables of Magnetic Structures n⁰ 4, Institute of Nuclear Techniques, Cracow (1970).
Meyer, C.: Thése de 3ᵉ cycle, Grenoble (1974).
Minkiwicz, V. J., Cox, D. E., Shirane, G.: (1970) Solid State Commun. *8*, 1001 (1970).
Misetich, A., Buch, T.: J. Chem. Phys. *41*, 2524 (1964).
Nathans, R., Will, G., Cox, D. E.: Proceedings of the International Conference on Magnetism, Nottingham (1964).
Ôno, K., Ito, A., Hirahara, E.: J. Phys. Soc. Japan *17*, 10, 1615 (1962).
— — *Fujita, T.:* J. Phys. Soc. Japan *19*, 2119 (1964).
O'Reilly, D. E., Tsang, T.: J. Chem. Phys. *40*, 734 (1964).
Owen, J., Thornley, J. H. M.: Rept. Progr. Phys. *29*, 675 (1966).
Patrie, M., Chevalier, R.: Compt. Rend. *263*, 1061—1064 (1966).
Pauling, L.: The nature of the chemical bond. Ithaca, N. Y.: Cornell Univ. Press 1960.
Pernet, M., Quezel, G., Coing—Boyat, J., Bertaut, E. F.: Bull. Soc. Franc. Mineral. Crist. *92*, 264 (1969).
Plumier, R.: Thesis, Paris (1968).
Rimmer, D. E.: Proc. Int. Conf. on Magnetism, Nottingham (1964).
Romanov, V. P., Chechershaya, L. F.: (1972) Phys. Stat. Sol. 49b, K 183 (1972).
Schoemaker, V., Stevenson, D. P.: J. Am. Chem. Soc. *63*, 37 (1941).
Shannon, R. D.: Chem. Commun. *1971*, 881.
— *Calvo, C.:* J. Solid State Chem. *6*, 538 (1973).
— *Prewitt, C. T.:* Acta Cryst. *825*, 925 (1969).
— — J. Inorg. Nucl. Chem. *32*, 1427—1441 (1970).
Stevens, K. W.: Proc. Roy. Soc. (London *A 219*, 542 (1953).
Tarte, P.: personal communication (1973).

Thornley, J. H. M.: Thesis, Oxford (1962).
Vincent, H., Bertaut, E. F.: J. Phys. Chem. Solids *34*, 151 (1973).
— *Perrault, G.:* Bull. Soc. Franc. Mineral. Crist. *94*, 551—55 (1971).
Walker, L. R., Wertheim, G. K., Jacarino, V.: Phys. Rev. Letters *6*, 98 (1961).
Wells, A. F.: J. Chem. Soc. *1949*, 55.
Wertheim, G. K., Buchanan, D. N. E.: Phys. Rev. *161*, 478 (1967).
Windsor, C. G., Thornley, J. H. N., Griffiths, J. H. E., Owen, J.: (1962) Proc. Phys. Soc. *80*, 803 (1962).
Wyckoff, R. W. G.: (1960) Crystal structures, Vols. I—IV. New York: Wiley 1960.

Received June 4, 1974

Considerations on the Valence Concept

A. Kjekshus and **T. Rakke**

Department of Chemistry, University of Oslo, Blindern, Oslo 3, Norway

Table of Contents

I. Introduction

The concept of valence has been subject to revision over the years. Initially, valence was regarded as the combining power of an element and was derived from the composition of compounds. At the end of the period before the age of quantum chemistry, valence was generally formulated in relation to the octet rule (*1—3*), a simple relation which still finds useful application in modern chemistry.

In quantum-chemical treatments, it is necessary to introduce numerical approximations even for the simplest possible molecule, H_2^+. Among the schemes of approximations which have emerged, the valence bond (VB) and molecular orbital (MO) methods are most prominent. However, attempts to improve the shortcomings of simple VB or MO concepts in the treatment of real molecules seem to have blurred the once clear-cut definition of valence.

The concept of the localized electron-pair bond (*i.e.* one electron pair allotted to each bond and vice versa) from the simple *Lewis* (*2*) theory was given quantum mechanical justification by *London* (*4*). In all essentials, the localized electron-pair bond has maintained an unshaken position in large areas of modern quantum chemistry. Ultimately, the attitude towards the fundamental nature of the electron-pair bond is a matter of faith, but even the most resolute believer must admit that serious shortcomings arise from this postulate (see below). Even though there are relatively few apparent exceptions for small and medium-sized molecules, numerous exceptions are found when attention is directed toward the macromolecular type which prevails in most inorganic solids. Attempts to remedy this unsatisfactory situation include the *Slater-Pauling* (*5—8*) and *Hund-Mulliken* (*9—16*) theories, where the requirement that a localized electron pair be associated with each bond is dropped. In spite of the amount of effort already devoted to problems in this field, we think a critical reexamination of the valence concept will be worth-while. Since quantum chemistry is far from its final goal of predicting the nature of multicomponent systems from calculations based on first principles, the most useful way of approaching the problems is through generalization from simple systems.

Because of the apparent success frequently achieved by applying the simple octet rule, we took this rule as our starting point. However, the plain formulation of the octet rule is somewhat cumbersome and is conveniently substituted by the "generalized" (8-N) rule (*17—28*). The generalized rule has received considerable attention in recent years as a powerful tool for predicting compound semiconductors.

II. The Generalized (8-N) Rule

Most current versions of the generalized (8-N) rule differ in notation or in their treatment of non-bonding electrons. Although the lack of a consistent notation causes some confusion, formal problems of this type will not be considered here. Any one scheme of notation seems just as good as the others but, for convenience, that previously used by one of the present authors (*22*) is adopted with some necessary modifications.

The most common formulation of the generalized (8-N) rule states that virtually all compound semiconductors

$$C^{1}_{c_1}\,C^{2}_{c_2}\cdots\cdots C^{j}_{c_j}\,A^{1}_{a_1}\,A^{2}_{a_2}\cdots\cdots A^{i}_{a_i} \qquad \text{(Def. II.1)}$$

(abbreviated to C_cA_a whenever precise definition is unnecessary; C = cationic element and A = anionic element, *i.e.* the more electropositive and the more electronegative element, respectively)

obey the relation

$$n + P - Q = 8a \qquad \text{(Eqn. II.1)}$$

where, per formula unit,

n is the total number of valence electrons, (Def. II.2)

P is the total number of electrons involved in A—A bonds, (Def. II.3)

Q is the total number of electrons involved in C—C bonds, and (Def. II.4)

$a = \sum_i a_i$ is the total number of anionic constituents. (Def. II.5)

This rule, which is an extension of the (8-N) rule for the elements (*17*), was developed on an empirical basis (*18—21*) and later shown (*22*) to be a mathematical formulation of the requirement that there be complete octets on all A's. Moreover, the rule can be extended (*22*) to allow configurations other than octets:

$$n + P - Q = \sum_i N_i a_i \qquad \text{(Eqn. II.2)}$$

where

N_i is the number of electrons required to fill the valence shell of an atom A^i. (Def. II.6)

A semi-theoretical justification of Eqn. II.1 has been presented independently by *Hulliger* and *Mooser* (*23*).

1. Relation to Classic Valence

The earlier derivation (*22*) of Eqn. II.2 was based on classic valence concepts, although these had clearly been influenced by quantum-chemical ideas. Even within the classic approach, a critical reexaimination of the underlying assumptions is valuable in order to establish what is the minimum number needed to deduce Eqn. II.2.

The first assumption implicitly introduced (Def. II.1) was:

The constituents of the compound can be distinguished as either anionic or cationic. (Hyp. II.1)

Valences x_i and y_j for the two kinds of constituents were introduced by taking

$(N_i - x_i)$ as the number of electrons in the valence shell of A^i, and (Def. II.7)

y_j as the number of electrons from the valence shell of C^j in-volved in bonding. (Def. II.8)

Obviously, these definitions need clarification. Chemists commonly assume $N_i = 8$, which allows $(N_i - x_i)$ to be interpreted as the group number of A^i in the periodic system; x_i (fixed for each group) can then be interpreted as follows:

(i) x_i is the number of "valence bars" (*i.e.* electron-pair bonds) formed by A^i for simple molecules. Whenever x_i appeared to exceed the number of near neighbours around A^i, multiple bonds had to be assumed.

(ii) x_i is the number of electrons required by A^i to form an anion for macromolecular solids.

Where both interpretations were found to be inappropriate, the classic valence concept was reformulated, either by taking a different value for N_i (*e.g.* the idea of expanded octets) or by reclassifying certain A^i atoms as cationic constituents in particular cases.

The classic valence for cationic elements has also been associated with the position of C^j in the periodic system and interpreted (analogous to (i) and (ii)) as:

(i') y_j is the number of "valence bars" from C^j to near neighbours for simple molecules.

(ii') y_j is the number of electrons surrendered by C^j to produce a cation for macromolecular solids.

The concepts (i) and (i′) or (ii) and (ii′) together give rise to the classic rule

$$\sum_i a_i x_i = \sum_j c_j y_j \,, \qquad \text{(Eqn. II.3)}$$

which expresses saturation of the valences of A^i with respect to C and of C^j with respect to A. In order to allow for the realistic possibilities of A—A and/or C—C bonding, additional specifications are introduced within the framework of classic valence:

p_i and q_j are the number of A—A bonds formed per A^i and C—C bonds per C^j, respectively. (Def. II.9)

Eqn. II.3 is then generalized to

$$\sum_i a_i(x_i - p_i) = \sum_j c_j(y_j - q_j) \,. \qquad \text{(Eqn. II.4)}$$

This equation can be said to represent the condition of complete saturation of all *predetermined* (in relation to the periodic system) anionic and cationic valences. There are, however, numerous examples of compounds whose predetermined classic valences do not satisfy Eqn. II.4. Although these inconsistencies could, in principle, have been cured in several ways, chemists have traditionally got round the problem by maintaining the anionic valences, and leaving the adjustable cationic valences to be determined from Eqn. II.4 or equivalents thereof. It follows that Eqn. II.4 can no longer be seen as an expression having general significance for required saturation of all valences, since it now merely expresses the already invoked saturation of anionic valences. There are many cases where it is not even sufficient to manipulate the cationic valences. Therefore, the apparent symmetry of Eqn. II.4 does not represent a basic chemical principle.

In the original (*22*) derivation of Eqn. II.2, Eqn. II.4 was, in effect, used as the starting point for the discussion; the logic was accordingly obscured by the fact that Eqn. II.4 already contained all the essentials of the object of derivation. Although these deductions are improper in the classic sense, it is still possible that Eqn. II.4 may be admissible in an extended valence scheme. However, in this situation fundamental difficulties arise in correlating the number of bonds p_i and q_j (Def. II.9) with the total number of electrons (Defs. II.3 and II.4). Hence the relations

$$P = \sum_i a_i p_i \qquad \text{(Eqn. II.5)}$$

and

$$Q = \sum_j c_j q_j \qquad \text{(Eqn. II.6)}$$

are non-trivial without additional assumptions.

In order to establish Eqn. II.2, it is in fact unnecessary to introduce valences defined according to Defs. II.7 and II.8. The key parameters are

g_i, the number of electrons in the valence shell of A^i, and (Def. II.10)

p_i', the number of electrons furnished per A^i to A—A bonds, and (Def. II.11)

N_i from (Def. II.6)

Attention is thus focused on the anionic constituents; an analogous dissection of the cationic situation is not required at this stage and it is sufficient to lump the valence features of the cationic constituents together in the gross term

Y, which specifies the total number of electrons involved in bonding from the valence shells of all C atoms. (Def. II.12)

For further progress, we assume that

The actual number of electrons involved in A—A bonds per A^i is $2p_i'$, (Hyp. II.2)

i.e. A^i and its surrounding A atoms contribute an equal number of electrons to the A^i—A bonds, fractional as well as electron-pair bonds being permitted. According to Defs. II.10 and II.11 and Hyp. II.2, the presence of A—A bonds increases the apparent number of electrons in the valence shell of A^i to $(g_i + p_i')$. When the A—C bonds are taken into account as well, the apparent number of electrons in the valence shell of A^i is further increased to, say, N_i'. The requirement that

The valence shell of A^i is completely filled with electrons (Hyp. II.3)

means that $N_i' = N_i$ according to Def. II.6. Hence, on the introduction of Q from Def. II.4 and Y from Def. II.12, the relation

$$\sum_i a_i (g_i + p_i') + Y - Q = \sum_i a_i N_i \qquad \text{(Eqn. II.7)}$$

is substantiated. Identification of

$$n = \sum_i a_i g_i + Y \qquad \text{(Eqn. II.8)}$$

according to Defs. II.2, II.10, and II.12, and the trivial equalization of $\sum_i a_i p_i'$ and P (Def. II.3) show that Eqn. II.7 is equivalent to Eqn. II.2, which was the object of the derivation.

The above argument requires no detailed knowledge of the nature of the A—C bonds, the only requirement being the rather vague assumption in Hyp. II.1. The nature of the C—C bonds is completely unspecified, and the A—A bonds are only superficially specified through Hyp. II.2. With so few assumptions, it is not surprising that numerous compounds of various categories comply with Eqn. II.2. However, because it is formulated in terms of overall quantities, Eqn. II.2 can provide tests and predictions of only limited significance; this applies in particular to the cationic constituents. It is not possible to remedy this situation within the classic or quasi-classic framework.

For reasons of symmetry it is tempting to introduce

> q_j' as the number of electrons furnished per C^j to C—C bonds, (Def. II.13)

and the subsequent trivial equalization of $\sum_j c_j q_j'$ and Q (Def. II.4), in the hope of obtaining a more varied picture of the bonding situation for the cationic constituents. Although clearly facilitating the use of Eqn. II.2, this type of notational manipulation is rather sterile. A further aspect of the classic valences is evident through the combination of Defs. II.7 and II.10, and Defs. II.8 and II.12 to obtain the trivial relations $g_i = N_i - x_i$ and $Y = \sum_j c_j y_j$, respectively. These, on substitution into Eqn. II.7, give

$$\sum_i a_i (x_i - p_i') = \sum_j c_j (y_j - q_j') , \qquad \text{(Eqn. II.9)}$$

which is the analogue of Eqn. II.4 in terms of electron numbers instead of bonds. The symmetry of Eqn. II.9 is obviously a consequence of the asymmetric valence definitions (Defs. II.7 and II.8). This symmetry/asymmetry problem is a characteristic feature of the classic or quasi-classic valence scheme; it can be shifted from one definition or relation to another, but, as a matter of course, it cannot be removed. As will become evident from the following sections, the question of symmetry versus asymmetry in connection with valence is, in reality, somewhat artificial.

2. Relation to Simple Molecular Orbital Language

In endeavours to modernize the valence concept it appears all too easy either to take over unproductive ideas from the classic treatment or, to reject useful concepts. We must therefore make detailed analyses of definitions and assumptions as we did in the preceeding section. Without apparent loss of generality, it is convenient to maintain our goal of clarifying the physical background and implications of the generalized (8-N) rule as a framework around the continued considerations of valence. When departing from the intuitive classic scheme, it is especially important to make the rules of the game explicit and to emphasize any approximation and/or shortcoming inherent in the general model. The rules accepted in this section are those of the MO theory based on the LCAO approximation.

The classic case distinguishes between an atomic core, which is essentially unperturbed by bonding, and a valence shell whose content may be accessible to bond formation. Since we suppose this simplifying assumption to be maintained in the MO treatment, an atomic orbital belonging to the valence shell will be termed a valence atomic orbital (VAO). For the construction of MOs, we utilize the following general results of the MO/LCAO model:

The resulting number of MOs must equal the number of VAOs used in their construction. (Thm. II.1)

MOs can be classified according to their general bonding character as bonding, non-bonding, and anti-bonding. (Thm. II.2)

Moreover, it is assumed that the loosely defined distinction between A and C (Def. II.1) can be maintained and is somewhat further specified through

M_i, the number of VAOs per A^i, and (Def. II.14)

r_i, the number of VAOs per A^i involved in the construction of A^i—C MOs. (Def. II.15)

This leads to the assumption that

$\sum_i a_i r_i$ is the total number of *bonding* A—C MOs per formula unit. (Hyp. II.4)

By analogy with Def. II.15,

Let r_j' be the number of VAOs per C^j involved in the construction of C^j—A MOs. (Def. II.16)

However, it is superfluous to formulate the analogue of Def. II.14. From Defs. II.15 and II.16, and Hyp. II.4, the "conservation of orbitals" through Thms. II.1 and II.2 gives the result:

The total number of *anti-bonding* A—C MOs per formula unit is $\sum_j c_j r_j'$. (Cor. II.1)

Hyp. II.4 and the consequent Cor. II.1 merely generalize the intuitively acceptable idea that bonding states are associated with anions and anti-bonding states with cations. However, the energetic interpretation may be misleading and must certainly not be projected back into Hyp. II.4 and Cor. II.1, which concern only the number, not the kind of VAOs involved. In counting bonding and anti-bonding MOs it is commonly assumed that the total numbers of each are equal; nevertheless, the possibility that $\sum_i a_i r_i \neq \sum_j c_j r_j'$ is not excluded in the following discussion.

It is also necessary to provide notational specifications of the A—A bonds. We take for the present purpose

p_i as the number of VAOs per A^i involved in the construction of A^i—A MOs, and (Def. II.17)

$\nu \sum_i a_i p_i$ as the total number of *bonding* A—A MOs per formula unit. (Def. II.18)

Then, from Thms. II.1 and II.2,

The total number of *anti-bonding* A—A MOs per formula unit is $(1-\nu) \sum_i a_i p_i$. (Cor. II.2)

For bonds between the same kind of atoms (A^i—A^i bonds) $\nu = \frac{1}{2}$ is the most natural assignment, but for bonds between A^i and other A atoms, $\nu \neq \frac{1}{2}$ cannot be ruled out. Hence

$$0 \leqq \nu < 1. \qquad \text{(Cor. II.3)}$$

Of the M_i (Def. II.14) VAOs available per A^i, the number $(r_i + p_i)$ is according to Defs. II.15 and II.17 used for the construction of bonding and anti-bonding MOs. Therefore

$(M_i - (r_i + p_i))$ is the number of *non-bonding* MOs associated with A^i. (Cor. II.4)

This completes the amount of pure MO formalism required at this stage, but the empty MO skeleton must be related to the electrons occupying the MOs. The common assignment:

Each MO can accommodate up to two electrons (Thm. II.3)

is chosen for the present purpose, where interest is focused on the counting of orbitals or electrons and not on their energy distribution. We could equally well have chosen to use different MOs for electrons of different spin, which intuitively is a very satisfactory approach (*29*), accounting inter alia for most of the electron correlation energy. The correspondence between number of MOs and number of electrons will in this treatment be designated by the following notations defined per formula unit:

$R^0 = \varrho \sum_i a_i r_i$ and $R^* = \varrho^* \sum_j c_j r_j'$ are the total numbers of *bonding* and *anti-bonding* *A—C* electrons, respectively. (Def. II.19)

$P^0 = \pi v \sum_i a_i p_i$ and $P^* = \pi^* (1 - v) \sum_i a_i p_i$ are the total numbers of *bonding* and *anti-bonding* *A—A* electrons, respectively. (Def. II.20)

$U = \mu \sum_i a_i (M_i - (r_i + p_i))$ is the total number of *non-bonding* electrons associated with all *A* atoms. (Def. II.21)

According to Defs. II.18—II.21, Hyp. II.4, and Cors. II.1, II.2, and II.4, the parameters ϱ, ϱ^*, π, π^*, and μ can be interpreted as the *average occupation numbers* per MO in question. Hence, from Thm. II.3,

$$0 \leq \varrho, \varrho^*, \pi, \pi^*, \mu \leq 2. \quad \text{(Cor. II.5)}$$

The analogous assumption that Q (Def. II.4), in general, is composed of bonding (Q^0) and anti-bonding (Q^*) C—C electrons is also permitted, but need not be considered here. Moreover, the analogue of Def. II.21 (still undetailed) is contained in the gross term specification:

T is the total number of non-bonding electrons associated with all C atoms per formula unit. (Def. II.22)

In the classic approach it was natural to interpret n according to Eqn. II.8, but from the MO point of view it is more natural to identify n as follows:

$$n = R^0 + P + U + Q + T + R^* \qquad \text{(Eqn. II.10)}$$

where the gross term symbols are as specified in Defs. II.3, II.4, II.19, II.21, and II.22. In this equation $P = P^0 + P^*$ and $Q = Q^0 + Q^*$ include both bonding and anti-bonding electrons. Non-bonding electrons were not explicitly considered in the classic case, but U (Def. II.21) was implicitly included in n whereas the possibility that $T \neq 0$ (Def. II.22) was automatically eliminated. This distinction (asymmetry) in the treatment of non-bonding A and C electrons arises out of the classic valence definitions (Defs. II.7 and II.8). R^* is the only term in Eqn. II.10 which has no "classic analogue" or cannot be included in a classic valence definition through reformulation. It is in the main the occurrence of this term that justifies a reexamination of the concept of valence (see below).

According to Defs. II.6 and II.14, and Thm. II.3,

$$M_t = N_t/2. \qquad \text{(Eqn. II.11)}$$

The elimination of U in Eqn. II.10 by way of Def. II.21, and the use of Defs. II.19 and II.20 (and the trivial equality $P = P^0 + P^*$) to identify terms, demonstrate the equivalence with

$$n + \left(\frac{\mu}{\varrho} - 1\right) R^0 + \left(\frac{\mu}{\pi\nu + \pi^*(1-\nu)} - 1\right) P - (Q + T + R^*)$$
$$= \frac{\mu}{2} \sum_t a_t N_t, \qquad \text{(Eqn. II.12)}$$

where N_t replaces M_t according to Eqn. II.11. When we examine the basis of Eqn. II.12, its fairly general validity becomes evident; more specialized equations can be obtained by the introduction of additional limiting assumptions.

Among the many special cases of Eqn. II.12 open to consideration, the task of relating it to Eqn. II.2 has a particular interest. In the attempt to achieve this goal, T (Def. II.22) may be eliminated through a redefinition of n. On the purely formal level one can also remove R^* (Def. II.19) in the same way, although it is obvious that bonding and anti-bonding electrons may not be subjected to the same manipulations as the non-bonding electrons. Thus, the only acceptable way of removing R^* is to make

$$R^* = 0, \qquad \text{(Eqn. II.13)}$$

which implies that there are no anti-bonding C—A electrons. From this point, conformity between Eqns. II.2 and II.12 is obtained by making the specializations

$$\mu = 2, \qquad \text{(Eqn. II.14)}$$

$$1 - \frac{\mu}{\varrho} = 0, \text{ and} \qquad \text{(Eqn. II.15)}$$

$$\frac{\mu}{\pi\nu + \pi^*(1-\nu)} - 1 = 1. \qquad \text{(Eqn. II.16)}$$

Some interpretation of these equations is called for. According to Cor. II.4, Thm. II.3, and Def. II.21, the implication of Eqn. II.14 is that *all non-bonding A levels are completely filled.* The use of Hyp. II.4, Thm. II.3, Def. II.19, and Eqn. II.14 reveals that Eqn. II.15 expresses the *complete filling of all C—A bonding levels.* On the introduction of Eqn. II.14, Eqn. II.16 is rearranged to

$$\pi\nu + \pi^*(1-\nu) = 1. \qquad \text{(Eqn. II.17)}$$

An immediate consequence of Eqn. II.17 is that Eqn. II.5 is also valid when p_t represents number of VAOs (Def. II.17) rather than number of A^t—A bonds (Def. II.9). This limitation implies that the *total number of electrons* involved in A—A bonds (Def. II.3) equals the *total number of VAOs* used in the construction of the A—A MOs (Def. II.17). The discussion of Eqn. II.17 can be continued along one of two lines, depending on the value of π^*.

(i) $\pi^* = 0$ means according to Def. II.20 that all the anti-bonding A—A levels are empty. Eqn. II.17 is reduced to $\pi\nu = 1$ which, from the general constraints of Cors. II.3 and II.5, leads to $1 < \pi \leqq 2$ and $\frac{1}{2} \leqq \nu < 1$. It must be recognized that the commonly accepted condition $\nu = \frac{1}{2}$ is only one of several possibilities, which, according to $\pi = 2$, states that all the bonding A—A levels are completely filled.

(ii) $\pi^* \neq 0$. This condition has meaning only when A—A bonding occurs, in which case $P^0 > P^*$ and hence, from Def. II.20, $\pi\nu > \pi^*(1-\nu)$. The elimination of ν by the use of Eqn. II.17 gives $1 > (1-\pi^*)/(\pi-\pi^*) > \pi^*/(\pi+\pi^*)$. At first sight, it may appear tempting to put $\pi = 2$ whenever $\pi^* \neq 0$, but this is certainly not the only possibility. If we recall that π and π^* represent average occupation numbers, both parameters may take intermediate (Cor. II.5) values depending on the location of the various A^i—A bonding and anti-bonding levels. This situation may give rise to unpaired electron configurations. A detailed discussion of the implications which follow from $\pi^* \neq 0$ is out of the question because of the almost infinite number of possibilities that may be imagined in the general case.

With the specializations represented by Eqns. II.13—II.16, the generalized (8-N) rule of Eqn. II.2 was obtained from Eqn. II.12, neglecting the non-bonding C electrons. Other equations intermediate between Eqn. II.12 and Eqn. II.2 could easily have been formulated by discarding one or more of the above specializations. If we consider the possibility of evaluating the parameters involved in Eqn. II.12 (see Section V), the gross term quantities n, R^0, P, Q, T, and R^* may be difficult to approach, but the average occupation numbers μ, ϱ, π, and π^* are even more inaccessible. Although the general form of Eqn. II.12 is thus of limited applicability, it has the advantage of providing the assurance that this really is the most general version of the (8-N) rule obtainable within the MO/LCAO framework. (The distinction between C and A through Hyp. II.4 is the only limitation imposed on the unspecialized MO/LCAO scheme in order to deduce Eqn. II.12.) Moreover, Eqn. II.12 is formulated in such a way as to permit explicit specializations, the physical implications of which can then be further appraised. (The situation that arises when the distinction between C and A is removed is discussed in Section III.)

From the MO point of view, Eqn. II.2 suffers from the weaknesses of neglecting *a priori* both the non-bonding C and the anti-bonding C—A electrons. For MO purposes, a more satisfactory equation would include these parameters, *viz.*

$$n + P - (Q + T + R^*) = \sum_i a_i N_i \qquad \text{(Eqn. II.18)}$$

However, for the purpose of testing or predicting the properties of specific compounds by way of Eqn. II.18, such improvements of the generalized (8-N) rule are somewhat artificial. Clearly, since no assumptions regarding the nature of Q and T are needed to establish Eqn. II.18, attempts to reverse the picture with the object of extracting decisive information about these parameters are ill-advised. Moreover, as we need to know only the valence situation of C with respect to A in order to derive Eqn. II.18, we are formally free to choose arbitrary values for Q and T, provided the same numbers are included in n. The desire to carry on with at least the term Q in Eqn. II.2 is obviously a legacy from the classic scheme, where the saturation of *predetermined* valences was the guiding principle (see II.1). If we put $n' = n - (Q + T)$, Eqn. II.18 reads

$$n' + P - R^* = \sum_i a_i N_i. \qquad \text{(Eqn. II.19)}$$

If the situation for Q and T is disappointing owing to their connections with C only, the position of the term R^* is intermediate between the two extremes: (*i*) the R^* electrons belong purely to C; (*ii*) the R^* electrons are shared equally between C and A. The former extreme is intuitively acceptable only if the C—A bonds are purely ionic, in which case the MO term anti-bonding loses much of its original significance. In (*ii*) and intermediate cases, the only excuse for excluding R^* from Eqn. II.19 would be explicitly to assume Eqn. II.13.

Before we close this MO/LCAO discussion of the generalized (8-N) rule, we note that a derivation of Eqn. II.1 has been reported by *Hulliger* and *Mooser* (*23*) on a similar basis. However, a careful analysis of their treatment reveals that, in addition to features of general MO/LCAO theory (Thms. II.1—II.3) and necessary assumptions (equivalents of Hyps. II.1—II.3), they also introduce some superfluous assumptions and specializations. This not only obscures the treatment, it also introduces new aspects which it may be instructive to dwell on in some detail. In order to keep the number of notational symbols to a minimum, the definitions already invoked in the preceding discussion will be utilized as far as possible. However, the disposition and layout of their paper differ significantly from ours; since, moreover, many of *Hulliger* and *Mooser's* assumptions are to be classified as being only partly superfluous, some quotations are inescapable.

To make a distinction between A and C (Def. II.1) seems at first sight intuitively obvious and this assumption was also made by *Hulliger* and *Mooser*. However, these authors introduce an unnecessary limitation:

> The molecular system is a crystalline solid, (Stm. II.1)

in order to facilitate the use of features from band theory (BT). This may appear to give certain advantages in that (*i*) the translational symmetry of the direct lattice and (*ii*) the potential within the unit cell are the deciding factors for the band structure. Hence, it is necessary to consider and discuss the Schrödinger equation only within the unit cell. Although this may be a relief, the apparent benefit comes at the expense of the complications associated with bands, which in the present context serve only to obscure the treatment.

Hulliger and *Mooser*, apparently in order to justify the use of a simple binary formula (A_aC_c) without imposing an unnecessary constraint, make the following claim:

> The energy differences between MOs having the same bonding character are relatively small compared to the energy differences between bonding and anti-bonding MOs. (Stm. II.2)

Although they do not say so explicitly, the phrase "same bonding character" is meaningless unless MOs are classified in the 8 categories: 6 representing $A—A$, $A—C$, and $C—C$ bonding and anti-bonding MOs, and 2 representing A and C non-bonding MOs. A closely associated suggestion is that each MO corresponds to a band of finite width and that

> all MOs of the same bonding character are representative of one composite band. (Stm. II.3)

Moreover, in accordance with *Wilson's* (*30*) criterion for semiconductivity, they state:

> The number of valence electrons in a semiconductor must equal twice the number of AOs contained in one or more of the composite bands. (Stm. II.4)

Stm. II.4, which may be regarded as a kind of MO—BT equivalence of Hyp. II.3, expresses the complete filling of levels. (Classically, only the A atoms had to be considered.) Following *Hulliger* and *Mooser:*

> The most common case is that not only the bonding, but also the non-bonding A MOs are occupied and, if each A has only C neighbours, and vice versa, the total number of occupied MOs is specialized to $4a$. (Stm. II.5)

The combined contents of Stms. II.4 and II.5 convey the MO—BT equivalence of the octet rule (Hyp. II.3 with Def. II.6 specialized to $N_t = 8$). If A—A and/or C—C bonding interactions occur, Stm. II.5 needs modification. To achieve this, p and q (Def. II.17 and its C—C analogue) are introduced through:

A number $\frac{1}{2}ap$ of A orbitals are split away to form anti-bonding MOs which remain empty; and (Stm. II.6)

A number $\frac{1}{2}cq$ of C orbitals are split away to form bonding MOs which become filled. (Stm. II.7)

According to the notations introduced in Defs. II.18 and II.20, Stm. II.6 expresses that $\nu = \frac{1}{2}$ and $\pi^* = 0$. The superfluity of Stm. II.7 becomes evident if we note that the corresponding parameters for the C—C bonds are not yet defined. (Anticipating the notational symbol introduced in Eqn. II.21, $\gamma = 2$.)

The combination of Stms. II.5—II.7 leads to the conclusion:

A number $(4a - \frac{1}{2}ap + \frac{1}{2}cq)$ of MOs is occupied. (Stm. II 8)

The non-bonding C MOs are not covered by any of Stms. II.1—II.8, but instead of a subsequent manipulation with n (Def. II.2), these can equally well be omitted at this stage. To achieve further progress, *Hulliger* and *Mooser* had to drop the distinction between occupation and filling, and in accordance with Stm. II.4 they assume:

All occupied MOs are completely filled in a semiconductor. (Stm. II.9)

On combining Stms. II.8 and II.9, they readily deduce Eqn. II.1.

It is evident from the above that the requirement for semiconductivity should also be added to the list of superfluous assumptions. However, the association of the generalized (8-N) rule with the question of semiconductivity is not in itself irrelevant. It is also worth noting that R^* (Def. II.19) does not enter into *Hulliger* and *Mooser's* treatment. The reason for this is hidden in Stms. II.3 and II.4, according to which $R^* = 0$. Surprisingly enough, *Hulliger* and *Mooser* do not draw this conclusion explicitly.

3. Relation between Normal and Inverted Versions

Up to now, most of the attention has, either explicitly or implicitly, been focused on the A components. This emphasis is more a result of the historical development of the (8-N) rule than of any particular primacy of the A atoms. In fact, the crucial distinction between A and C, based on excess versus deficiency in charge, does not enter into the derivation of Eqn. II.12 at all. It is therefore rather obvious that an expression analogous to Eqn. II.12 can be derived for the C components.

Eqn. II.12 was derived from Eqn. II.10 by the elimination of U (as explicit term) through Def. II.21. For this reason, Eqn. II.12 is not the most convenient form for the continuation where it is desirable to include U as the analogue of T (Def. II.22) as an explicit rather than implicit parameter. In addition to U, R^0 (Def. II.19), and P (Def. II.3) are the only quantities eligible for elimination from Eqn. II.10. Although U, R^0, and P are equally probable candidates, the rejection of the term P is inconvenient when what is desired is an explicit resemblance to Eqn. II.2. The advantage of removing the explicit occurrence of R^0 is that this parameter, as opposed to the two others, is a term common to the A and C components, giving rise to

$$n + (\alpha - 1)P + \left(\frac{\varrho}{\mu} - 1\right)U - (Q + T + R^*) = \varrho a M\,, \qquad \text{(Eqn. II.20)}$$

where a is introduced from Def. II.5 and

$M = (\sum_i a_i M_i)/a$ is the average number of VAOs per A, and (Def. II.23)

$\alpha = \varrho/(\pi\nu + \pi^*(1-\nu))$. (Def. II.24)

In order to arrive at an equation analogous to Eqn. II.20 with emphasis on the C constituents, the key parameter

M', expressing the average number of VAOs per C, (Def. II.25)

must be introduced. Moreover, the notational symbols

$c = \sum_j c_j$, the total number of C components, and (Def. II.26)

τ, the average number of electrons contained in each non-bonding C MO, (Def. II.27) are needed to establish

$$n + (\gamma - 1)Q + \left(\frac{\varrho}{\tau} - 1\right)T - (P + U + R^*) = \varrho c M' \,, \qquad \text{(Eqn. II.21)}$$

where the symbol γ is the analogue of α (Def. II.24) in Eqn. II.20. Hence, γ conveys information about the C—C interactions without requiring unnecessary (for the present purpose) specifications analogous to Defs. II.17, II.18, and II.20. Eqn. II.20 can be regarded the *normal*, hitherto most general version of the (8-N) rule; it is accordingly natural to regard Eqn. II.21 as its *inverted* version. Since Eqn. II.21 focuses attention on the C atoms, there is little point in maintaining P and U as explicit terms; the arguments for this are completely parallel to those presented in connection with the discussion of Eqn. II.12. However, when the A and C constituents are considered simultaneously, the freedom to manipulate n is limited since this parameter must have the same value in, say, both Eqn. II.20 and Eqn. II.21.

Before we continue the discussion of Eqns. II.20 and II.21, it is convenient to make some slight reformulations consequent on the introduction of

u and t as the average numbers of non-bonding MOs per A and C, respectively. (Def. II.28)

It follows from Defs. II.5, II.21, II,28, and Cor. II.4 that

$$u = U/\mu a \,, \qquad \text{(Eqn. II.22)}$$

and from Defs. II.22 and II.26 to II.28

$$t = T/\tau c \,. \qquad \text{(Eqn. II.23)}$$

Substitution of Eqns. II.22 and II.23 into Eqns. II.20 and II.21 gives

$$n = \varrho a M + (\mu - \varrho) a u + \tau c t + (1 - \alpha) P + Q + R^* \text{ and} \qquad \text{(Eqn. II.24)}$$

$$n = \varrho c M' + (\tau - \varrho) c t + \mu a u + (1 - \gamma) Q + P + R^* \,. \qquad \text{(Eqn. II.25)}$$

Of the parameters which occur in Eqns. II.24 and II.25, M and M' are in the particular position that they are of virtually atomic character. Furthermore, the initial assumptions concerning M and M' have indirectly a profound influence on other parameters in the two expressions. It is worth noting that through the selection of M_i (often 4) the idea that:

> Compounds that fulfil the generalized (8-N) rule can formally be ascribed an ionic formula (permitting *A—A* and/or *C—C* aggregates) and vice versa, (Stm. II.10)

is built into the model. By an appropriate specialization of N_i ($= 2\,M_i$), Stm. II.10 can more specifically be regarded as concerning Eqn. II.2. in which case classic ionic formulae are encountered. On application of the normal version of the generalized (8-N) rule, the implication of Stm. II.10 is that for the formal counting of electrons, the *A* atoms may be regarded as acceptors of all electrons involved in *C—A* interactions. (This extreme situation may be recognized as the *normal ionic formulation.*) The opposite situation arises in connection with the use of the inverted rule, where the *C* atoms hypothetically may acquire all *C—A* electrons (the *inverted ionic formulation*). In these processes, the *actual* *C—A* bonding state is left completely unspecified; it will be recalled that this is the characteristic feature of the generalized (8-N) rule formalism.

Certain features of the normal ionic formulation have been touched on by a number of authors (*e.g.* (*27, 31, 32*)) and very recently by *Parthé* (*28*). The basis of *Parthé's* scheme is the "iono-covalent hypothesis" where the bonding states of a given compound are regarded as a mixture of (hypothetical) limiting ionic and covalent states. Each of these bonding states is represented by mathematical expressions. Statements more or less similar to Stm. II.10 may have led *Parthé* to consider Eqn. II.1 as such an expression for the "ionic limiting bonding state". Although we fully agree with *Parthé* that an ionic formula is acceptable for the formal counting of electrons, it should be emphasized that the generalized (8-N) rule formalism does not allow deductions concerning the bonding state. (Despite the fact that the term "bonding state" is incorrect in this context, it is convenient to maintain it throughout the discussion of *Parthé's* scheme.)

It is easy to show that *Parthé's* mathematical expression(s) for the "covalent limiting bonding state" are not independent of Eqn. II.1 but, on the contrary, directly related to it. To substantiate this, one has merely to add Eqns. II.24 and II.25, rearrange to

$$2\,n = \varrho(aM + cM') + (2\,\mu - \varrho)au + (2\,\tau - \varrho)ct + (2 - \alpha)P + (2 - \gamma)Q + 2\,R^{*}, \qquad \text{(Eqn. II.26)}$$

and make some specializations. Consequent on the assumption of Eqn. II.1 (in conformity with *Parthé*), Eqns. II.13 to II.16 must be fulfilled as well as

$$M = 4 \,. \qquad \text{(Eqn. II.27)}$$

After these specializations, Eqn. II.26 reads:

$$n = (4a + M'c) + (au + ct) + (\tau - 2)ct + \tfrac{1}{2}(2 - \gamma)Q \,. \qquad \text{(Eqn. II.28)}$$

In order to obtain the expression(s) presented by *Parthé*, one has furthermore to make

$$\gamma = 2 \text{ or } Q = 0 \qquad \text{(Eqn. II.29)}$$

$$\tau = 2 \text{ or } t = 0 \qquad \text{(Eqn. II.30)}$$

$$M' = 4 \text{ or } 6 \qquad \text{(Eqn. II.31)}$$

in Eqn. II.28. It should be noted that the choice referred to in Eqn. II.31 concerns a distinction in the coordination of the C atoms (where *Parthé* makes the serious limitation of allowing only tetrahedral, square planar, and octahedral coordinations) as opposed to Eqns. II.29 and II.30, which are designed to eliminate the last two terms of Eqn. II.28. Since the specializations implied by Eqns. II.29 to II.31 are supplementary to those assumed in the derivations of Eqns. II.24 and II.25, *Parthé's* expression(s) for the "covalent limiting bonding state" are not independent of the generalized (8-N) rule formalism. It has already been substantiated that a characteristic feature of this formalism is that the actual *C—A* bonding state is left completely unspecified. Therefore, one obviously cannot combine Eqns. II.24 and II.25, which convey no information about a specific bonding state, so as to obtain a new expression (Eqn. II.26) that will carry such information. Clearly, any of Eqns. II.24 to II.26 can be used for the formal counting of electrons that are hypothetically distributed according to a covalent scheme. In this respect, hypothetical ionic and covalent bonding schemes represent equally valid modes of counting.

Closer inspection of the assumptions underlying the establishment of Eqn. II.20 (in particular Defs. II.15 and II.16 and Hyp. II.4) reveals that

$$\sum_i a_i r_i = \sum_j c_j r'_j \qquad \text{(Eqn. II.32)}$$

must be assumed in order to arrive at Eqn. II.21. As a consequence of Eqn. II.32, Hyp. II.4 cannot be regarded as a criterion for distinguishing between *A* and *C*. In fact, the necessity for such a distinction disappears, which implies that Def. II.1 and Hyp. II.1 are superfluous in the formalism. Another question is whether a realistic criterion does exist for a classification of A and C atoms in the general case. If we want to retain the intuitive classic content of the terms anionic and cationic, the most satisfactory way would be to determine the net effective charges on all atoms in a given compound. A full experimental or theoretical exploration of this problem is a pure wish at present, although it may be approached in particular and more extreme cases.

As an alternative, it is natural to try to design a criterion based on knowledge about energy-level diagrams for the electrons of the free atoms. If we neglect the atomic cores, we could as a first approximation consider the outer (valence) levels and characterize A and C atoms by such low- and high-lying levels, respectively. Applied on a quantitative basis, such an approach would yield an almost continuous scale wherein the location of an atom would depend on its position in the periodic system. At first sight, the problem may appear to be that of choosing the borderline between A and C elements. However, this is not the only problem because, wherever the borderline is drawn, the energy difference between, say, A^i and $A^{i'}$ can easily be greater than that between A^i and C^j. A rough interpretation of this situation is that the A^i—$A^{i'}$ bond is more ionic than the A^i—C^j bond. Consequently, this criterion also leads to meaningless classifications of the constituents in the general case (with reference to Def. II.1). The common use of the electronegativity concept to enforce a distinction between A and C atoms is clearly subject to the same weaknesses. (Questions concerning the significance of the various electronegativity scales do not enter into this conclusion.)

Some of the above difficulties can be overcome in certain cases, where a "guiding definition" of A may be atoms for which

$$M_i \text{ is a small number (usually 4) and} \qquad \text{(Stm. II.11)}$$
$$g_i \geqq M_i\,,$$

g_i being introduced through Def. II.10. As discussed in Section II.1, reclassification of certain A^i from anionic to cationic may be feasible when Stm. II.11 appears to break down. The arbitrariness of the latter type of manipulation clearly emphasizes that the desire to distinguish between A and C obscures the treatment rather than promotes progress. In conclusion, it may be recognized that the root of these difficulties is the requirement that three or more components of a given compound be grouped into *only* two categories, which inevitably is an oversimplification. In focusing attention on the somewhat artificial distinctions between A—A, A—C, and C—C bonds, one tends to overlook the essential question of whether there is or is not bonding interaction between two given atoms, regardless of their more or less arbitrary preclassification.

III. On the Valence Problem

In an attempt to relieve the difficulties associated with the classification of the constituents of a compound in only two categories, we relinquish the requirements of Def. II.1 and Hyp. II.1, assuming a general formula:

$B^1_{b_1}B^2_{b_2}\ldots.B^l_{b_l}$, where the superscript ($i = 1,2,\ldots., l$) refers to atoms which are chemically and structurally equivalent. (Def. III.1)

The essential novelty is that the number of classes have been increased from two to l (*i.e.* one class for each kind of atom). At this stage it might appear natural to explore the impact on the generalized (8-N) rule formalism consequent on the alterations introduced through Def. III.1. In order to achieve this, a number of notations more or less similar to those presented in Section II have to be introduced. Although some of these redefinitions will inevitably prove necessary, their number can be significantly reduced by postponing this discussion. A more important consideration is that, unless we examine the crucial problems concerning valence and bonding, we are likely to draw unproductive and even wrong conclusions.

Parallel with the change in general formula (Def. III.1), one is led to introduce

g_i, as the possible number of electrons available for bonding interactions per B^i, (Def. III.2)

which will be recognized as the *de facto* extension of Def. II.10 to all atoms. Def. III.2 is coupled to the periodic system via the recipe:

g_i is to be identified with the group number for atoms from IA, IIA, and IIIB–VIIIB, with the number of (outer) d electrons plus two for the transition elements, and the number of (outer) f electrons plus three for the lanthanides and actinides. (Stm. III.1)

Stm. III.1 merely serves to define the borderline between the atomic core and electrons which are accessible for bonding. In line with this, it is natural to introduce

v_i, as the actual number of electrons involved in bonding per B^i, (Def. III.3)

as a reformulation of the classic valences (*cf.* Defs. II.7 and II.8). Hence, g_i will be the maximum achievable valence for B^i, $v_i \leqq g_i$, the difference $(g_i - v_i)$ being acknowledged as the number of non-bonding electrons per B^i.

1. Considerations on Simple Bonding Principles

In retrospect, perhaps the most striking feature of Section II is that attention was focused on average and/or overall bonding parameters. Because of the alteration of the general viewpoint represented by Def. III.1, this is no longer a clear-cut advantage. On the contrary, a more realistic approach should evolve on turning to the particular $B^i - B^j$ bonds. The apprehension of the need for a more detailed account of these bonding interactions will in turn force us to dig deeper into the underlying general bonding principles. However, there appears to be no objection to maintain the MO/LCAO framework for this purpose.

The basic assumption of the MO/LCAO approach is that a wave function of the type

$$\begin{aligned} \Psi_{ij} &= \sum_{\sigma_{ij}} \lambda(\sigma_{ij})\psi(\sigma_{ij}) + \sum_{\sigma_{ji}} \lambda(\sigma_{ji})\psi(\sigma_{ji})\ ; \\ \psi(\sigma_{ij}) &= \sum_{\beta_i} \phi(\beta_i, \sigma_{ij}),\ \beta_i = 1,2,\ldots,b_i, \text{ and} \\ \sigma_{ij} &= 1,2,\ldots,s_{ij} \end{aligned} \qquad \text{(Eqn. III.1)}$$

describes the $B^i - B^j$ bonding interactions, where

s_{ij} is the number of AOs per B^i involved in $B^i - B^j$ interactions, (Def. III.4)

and $\phi(\beta_i, \sigma_{ij})$ is a representative AO wave function with the appropriate mixing coefficient $\lambda(\sigma_{ij})$. (According to Def. III.1, the mixing coefficients for wave functions $\phi(\beta_i, \sigma_{ij})$ with fixed σ_{ij} can be assumed equal.) Although no assumptions are needed concerning the AOs which enter into Eqn. III.1, it is customary to utilize one-electron wave functions for this purpose. This approach will implicitly also be assumed here, where details concerning energy levels, etc. are not aimed at. It should nevertheless be noted that the adoption of one-electron AOs is not free from problems, configurational interaction being only one special aspect encountered in this connection.

On the introduction of

S_{ij}^0 and S_{ij}^* as the numbers per formula unit of bonding and anti-bonding $B^i - B^j$ MOs, respectively, (Def. III.5)

the principle of Thm. II.1 concerning the "conservation of orbitals" can be expressed mathematically as

$$S_{ij}^0 + S_{ij}^* = b_i s_{ij} + b_j s_{ji} \text{ for } i \neq j, \text{ and} \qquad \text{(Thm. III.1)}$$
$$S_{ii}^0 + S_{ii}^* = b_i s_{ii} \,.$$

The second expression in Thm. III.1 is included in order to emphasize the special situation which prevails for $B^i - B^i$ interactions. Since all B^i atoms are fully equivalent (Def. III.1), the postulation of

$$S_{ii}^0 = S_{ii}^* \; (= \tfrac{1}{2} b_i s_{ii}) \qquad \text{(Thm. III.2)}$$

seems almost inescapable. When $i \neq j$ the situation becomes less clear-cut, although

$$S_{ij}^0 = S_{ij}^* \qquad \text{(Hyp. III.1)}$$

is nevertheless commonly accepted. Intimately connected with this is the assumption that

$$b_i s_{ij} = b_j s_{ji} \,, \qquad \text{(Hyp. III.2)}$$

which expresses the fact that equal numbers of AOs from B^i and B^j are involved in the $B^i - B^j$ interactions.

When attention is shifted from number of orbitals to number of electrons, the latter are traditionally introduced via coordination numbers. In line with this

Let k_{ij} be the (partial) coordination number of B^i with respect to B^j atoms, (Def. III.6)

and according to Def. III.1

$$b_i k_{ij} = b_j k_{ji} \,. \qquad \text{(Eqn. III.2)}$$

The commonly assumed correspondence between number of bonds and number of electrons can now be expressed as

$$b_i v_{ij} + b_j v_{ji} = 2\, b_i \theta_{ij} k_{ij}\,, \qquad \text{(Hyp. III.3)}$$

where

v_{ij} is the actual number of electrons involved in $B^i - B^j$ interactions per B^i, (Def. III.7)

and θ_{ij} (= 1, 2, or 3) is a factor which facilitates the (slight) generalization from single electron-pair bonds only, to double or triple bonds also. The assumption expressed by Hyp. III.3 may perhaps be adequate in cases where the total number $\sum_i b_i g_i$ (Def. III.2) of available electrons is sufficiently large, otherwise serious difficulties are encountered with this postulate.

Hyp. III.3 is in reality a mathematical formulation of the original electron-pair idea (*3*) without any explicit reference or connection to quantum chemistry. However, the electron-pair concept is built into both VB and MO theory through postulates of the types Hyps. III.1, III.2, and

$$s_{ij} = \theta_{ij} k_{ij}\,, \qquad \text{(Hyp. III.4)}$$

with the additional requirement that

S^0_{ij} and S^*_{ij} of the MOs are completely filled and completely empty, respectively. (Hyp. III.5)

This leads to an intimate connection between Hyps. III.1 to III.5, but it is worth noting that only circumstantial evidence justifies the exaltation of the electron-pair bond to a fundamental bonding principle. In fact, the bonding principles contained in Hyps. III.1 to III.5 are merely generalizations from quantum-mechanically based treatments of very simple systems. However, an analysis of some examples (H_2, C (diamond), various organic molecules, etc.) from this category reveals that a further bonding principle (although not recognized as such) also appears to be fulfilled. This is

$$b_i v_{ij} = b_j v_{ji}\,, \qquad \text{(Hyp. III.6)}$$

which is the analogue of Hyp. III.2 in terms of electrons rather than orbitals.

With respect to the simple molecular systems, Thms. III.1 and III.2, and Hyps. III.1 to III.6 all appear to be equally well satisfied. The process of generalization may invalidate this inference. As emphasized by the notations, Thms. III.1 and III.2 are considered to occupy a special position, and there are good reasons to believe that both can be raised to the status of fundamental bonding principles. However, the fundamental nature of Hyps. III.1 to III.6 must be judged on the basis of their applicability to actual compounds.

One of the many possible schemes of classification divides compounds according to assigned values of $\sum_i b_i g_i$.

(*i*) When $\sum_i b_i g_i \geqq \sum_i b_i(2\,k_i - k_{ii})$, where $k_i = \sum_j k_{ij}$ denotes the (total) coordination number of B^i, the correctness of Hyps. III.1 to III.5 is traditionally not questioned. A bond is termed "normal" when $\theta_{ij} = 1$ (Hyp. III.3), whereas (multiple) bonds corresponding to $\theta_{ij} = 2$ or 3 are regarded as being only somewhat less normal. The ionic cases (*i.e.* whenever a complete ionic description appears plausible) represent a subclass for which Hyps. III.1 to III.5 are essentially meaningless. An ionic bond is nevertheless traditionally also regarded as "normal".

(*ii*) When $\sum_i b_i g_i < \sum_i b_i(2\,k_i - k_{ii})$, Hyp. III.3 must obviously break down; such compounds are labelled electron-deficient. The resulting "abnormal" bonding situation is commonly characterized by the term fractional bonding. (If θ_{ij} is allowed to take values $0 \leqq \theta_{ij} \leqq 3$, Hyp. III.3 may only have to be rejected in part. The resonance concept of the VB theory is an alternative attempt to maintain essentially the idea behind Hyp. III.3.) A number of electron-deficient compounds are commonly placed in or forced into the ionic category, thus removing the label "abnormal".

Perhaps the most depressing fact associated with the consequences of the above division is the lack of consistency often found in treatments of compounds which are essentially isostructural. Take, for instance, the different descriptions of the bonding situation in B_2H_6 on the one hand, and the isostructural (*e.g.* Al_2Cl_6) molecules on the other: while the latter may be treated by the conventional bonding principles expressed in Hyps. III.1 to III.5, the treatment of the former (in terms of 3-centre bonds) breaks with Hyps. III.1 to III.4. A similar conclusion is in fact reached in the majority of "abnormal" cases. Other simple examples are provided by the alkali–metal hydrides (with NaCl-type structure), CuH (with ZnS-wurtzite type structure), etc. These examples are typical in that it is only when a scarcity of electrons and/or orbitals enforces a search for extraordinary bonding principles that Hyps. III.1 to III.4 are reluctantly (partly or completely) replaced by alternative characteristics. Although Hyps. III.1 to III.4 are entangled in a rather messy way, we have to find out which of them is the real cause of the trouble. The above examples, and a number of others, suggest that Hyp. III.3 is the root of the evil. Hence, Hyp. III.3 is discarded as an *a priori* fundamental bonding principle, although it is, of course, fully acceptable as a derived feature in particular cases.

The inherent difficulties associated with Hyps. III.1, III.2, and III.4 are of a somewhat different character for they concern orbitals rather than electrons. The importance of this distinction is that electrons are basically concrete, as opposed to the abstract concept of orbitals, which is moreover, meaningless unless imagined occupied by electrons. A complicating factor that enters into judgements about the usefulness of Hyps. III.1, III.2, and III.4, is the de facto endless number of remedies that can be introduced in order to raise the level of "sophistication" in MO treatments (choices of AO basis sets, spin–orbit couplings, configurational interactions, etc.). However, in the simpler descriptions based on MO/LCAO theory, too much attention is commonly paid to the empty MO skeleton and it tends to be forgotten that *electrons, not orbitals* are the subject of bonding interactions.

Now that we have rejected Hyps. III.1 to III.4 as fundamental bonding principles, it is rewarding to find that the effort put into the dissection of examples has given rise to a substitute, Hyp. III.6. The significance of Hyp. III.6 has perhaps been implicitly accepted for a long time, but a change in traditional chemical thinking appears to be needed before it is recognized as a fundamental bonding principle. Hence, to motivate the reader, we continue with a brief discussion of covalency and covalent bonds.

2. Covalent and Homopolar versus Ionic Bonds

As emphasized in Section II.3, the representation of bonding states by mathematical expressions of the type of Eqns. II.24 and II.26 may lead to confusion. The common use of the term "covalent" is another source of confusion. This term, originally introduced by *Langmuir* (*3*), distinguishes a hypothetical bonding state which fulfils Hyp. III.3 (with $\theta_{ij} = 1$) and the additional requirement that

The electrons are shared equally between the bonded atoms. (Stm. III.2)

While this definition of electron-pair "covalency" neglects the effective charges on the bonded atoms, the designation "homopolar" bond appears to be reserved for the special case of an electron-pair bond between uncharged atoms. However, covalency has intuitively become accepted as the opposite to ionicity.

The polarity of the various $B^i - B^j$ MOs may be defined in terms of the parameters $\lambda(\sigma_{ij})$ of Eqn. III.1, which for the present purpose is condensed and somewhat reformulated to

$$\Psi_{ij} = K(b_i s_{ij} \lambda_{ij} \psi_{ij} + b_j s_{ji} \lambda_{ji} \psi_{ji}) \qquad \text{(Eqn. III.3)}$$

The relative magnitudes of the resulting average mixing coefficients λ_{ij} and λ_{ji} per B^i and B^j average AO, respectively, give a measure of the polarity of the $B^i - B^j$ MOs. In simple cases (say, HCl) the concepts of orbital polarity and electron polarity (as a measure of ionicity) coincide. The extension of the latter inference to more complicated molecular systems is far from trivial, although such a conclusion is commonly accepted for qualitative purposes.

There are two extreme possibilities for the mixing coefficients in Eqn. III.3 $\lambda_{ij} \to 0$ and $\lambda_{ji} \to \infty$; $\lambda_{ij} \to \infty$ and $\lambda_{ji} \to 0$ both of which represent pure ionic bonding. Our hesitation concerning the interpretation of λ_{ij} and λ_{ji} in the general case does not invalidate this conclusion. Unfortunately, the pure ionic bonding situation is never realized in practice, where the mixing coefficients take intermediate values.

The idea of Stm. III.2 is intuitively basic in characterizing a hypothetical covalent bonding state. Moreover, the idea of Stm. III.2 is commonly coupled to the MO language via $\lambda_{ij} = \lambda_{ji}$ which, together with Hyps. III.1, III.2, III.4, and III.5 (specialized to $\Theta_{ij} = 1$), constitutes an altenative specification of electron-pair "covalency".

For, say, the AsH_3 molecule, the two extreme hypothetical ionic bonding states $As^{3+}\,3H^-$ and $As^{3-}\,3H^+$ are both in some respects acceptable. If the same six electrons are distributed in accordance with the requirement of Stm. III.2, we obtain a bonding state characterized by electron pairs and effectively neutral atoms. For this example, Hyps. III.1 to III.6 all appear to be satisfied. In fact, this is so with the majority of small and medium-sized molecules. Hence, the intuitive apprehension of covalency in contrast to ionicity is here automatically consistent with the idea of electron-pair "covalency".

Turning to macromolecular inorganic compounds, say ZnS, the two hypothetical ionic extremes are $Zn^{2+}S^{2-}$ and $Zn^{6-}S^{6+}$ (an inverted, unusual formulation). We can imagine a continuous array of possible electron distributions between these extreme limits, one of which is the electron-pair "covalent" bonding state. The association of covalency with $\lambda_{ij} = \lambda_{ji}$ in Eqn. III.3 warrants non-polar formal MOs. However, a different situation arises when electrons are permitted to enter the empty MO skeleton. The electron-pair "covalent" state corresponds to an even distribution of eight ($v_{Zn} + v_S = 2 + 6$) bonding electrons per formula unit between the tetrahedrally coordinated Zn and S atoms. Hence, filling of the four *non-polar* formal MOs results paradoxically in a well-defined *electron polarity* characterized by the ionic formula Zn^{2-}–S^{2+}. A closely related consequence following from the stipulation of electron-pair "covalency" as ideal (100%) covalency, has been drawn by *Goodman* (*33*), who points out that a hypothetical bonding state for ZnS composed of uncharged atoms is 50% ionic. *Goodman's* intuitively strange conclusions have provoked discussion (*34–36*), but this unfortunately has only contributed to obscure the real problem.

All the inconsistencies can be traced back to the often incompatible requirements of electron-pair "covalency" as the manifestation of the ideal covalent bonding state on the one hand, and covalency as the opposite to ionicity on the other. On the basis of the ideas advocated in Section III.1, it is comparatively easy to choose between the two alternatives and give the definition:

> The *complete covalent* bonding state for the B^i—B^j interactions is a hypothetical state in which the relevant electrons $b_i v_{ij} + b_j v_{ji}$ are equally shared and where $b_i v_{ij} = b_j v_{ji}$, (Def. III.8)

where v_{ij} is introduced from Def. III.7. Def. III.8, which will be recognized as combining the contents of Hyp. III.6 and Stm. III.2, is fully compatible with the common statement that the quantum-mechanical exchange effect is the fundamental characteristic of covalent bonding. The completely covalent case is governed by an uninhibited exchange of electrons, which, it will be recalled, is tantamount to equal electron sharing. Moreover, as is evident from Def. III.8, effectively neutral atoms are obtained and the intuitive idea that covalency is the opposite to ionicity is ensured, thus invalidating any distinction between the terms homopolar and covalent.

On passing from a completely covalent to progressively more electrovalent B^i—B^j bonding situations, the complete exchange is gradually inhibited because the electrons are spending more time in the neighbourhood of one (say B^j) of the atoms. If we put $\lambda_{ij} = v_{ij}\lambda'_{ij}/s_{ij}$ in Eqn. III.3, the expression can be rewritten as

$$\Psi_{ij} = K(b_i v_{ij} \lambda'_{ij} \psi_{ij} + b_j v_{ji} \lambda'_{ji} \psi_{ji}). \qquad \text{(Eqn. III.4)}$$

The average mixing coefficients λ'_{ij} and λ'_{ji} may now be interpreted as the (relative) average weights on electrons from B^i and B^j, respectively, which is to acknowledge that the physical significance of orbital mixing in fact stems from the exchange of electrons.

IV. The Neutral-Bonded Formalism

According to Section II, the strength and weakness of the generalized (8-N) rule formalism rest mainly on (i) the division of the compound constituents into only two categories and (ii) the use of overall bonding parameters. Explicitly or implicitly, this formalism is also influenced by Hyps. III.1 to III.4, which we rejected as fundamental bonding principles in Section III. The alteration in viewpoint consequent on the acceptance of Def. III.1 implies that each kind of atom (l in all) must be considered within the generalized (8-N) rule formalism. The reformulation imposed by this alteration is conveniently carried out in parallel with a switch to Thms. II.1 to II.3, III.1, and III.2 as the only fundamental MO bonding principles.

The redefined key parameter (cf. Defs. II.14, II.23, and II.25) in the generalized (8-N) rule formalism is

m_i, the number of AOs potentially available for bonding interactions per B^i. (Def. IV.1)

The definition:

u_i is the number of non-bonding MOs per B^i, (Def. IV.2)

and the introduction of s_{ij} from Def. III.4 give

$$m_i = u_i + \sum_j s_{ij} \qquad \text{(Eqn. IV.1)}$$

The splitting of the B^i–B^j MOs into bonding and anti-bonding MOs is taken care of by the average parameter

$$v_i = \Big(\sum_{j \neq i} S_{ij}^0\Big) \Big/ \Big(\sum_{j \neq i} S_{ij}\Big), \qquad \text{(Def. IV.3)}$$

where S_{ij}^0 is introduced from Def. III.5 and (for $j \neq i$)

$$S_{ij} = b_i s_{ij} + b_j s_{ji}. \qquad \text{(Def. IV.4)}$$

The relation between orbitals and electrons is effected by the average parameters Δ_i, Δ_i^*, δ_i, δ_i^*, and μ_i as defined per formula unit through

$$R_i^0 = \Delta_i \nu_i \sum_{j \neq i} S_{ij} \quad \text{and} \qquad \text{(Def. IV.5)}$$

$$R_i^* = \Delta_i^* (1-\nu_i) \sum_{j \neq i} S_{ij},$$

which are the numbers of bonding and antibonding B^i—B^j electrons, respectively, for fixed i, but variable $j \neq i$,

$$P_i^0 = \tfrac{1}{2}\delta_i b_i s_{ii} \quad \text{and} \quad P_i^* = \tfrac{1}{2}\delta_i^* b_i s_{ii}, \qquad \text{(Def. IV.6)}$$

which are the numbers of bonding and antibonding B^i—B^i electrons, respectively, and

$\mu_i b_i u_i$, which is the number of non-bonding B^i electrons. (Def. IV.7)

The interpretation of Δ_i, Δ_i^*, δ_i, δ_i^*, and μ_i as average electron occupation numbers for the MOs in question follows from Defs. III.4, III.5, and IV.2 to IV.4, and Thms. III.1 and III.2. The total number of electrons n_i (per formula unit) associated with MOs involving AOs from B_i is, according to Defs. IV.5 to IV.7

$$n_i = \mu_i b_i u_i + R_i^0 + R_i^* + P_i^0 + P_i^*. \qquad \text{(Eqn. IV.2)}$$

By eliminating the explicit appearance of u_i through Eqn. IV.1, and introducing the parameter

$$\omega_i = (b_i \sum_{j \neq i} s_{ij})/(\sum_{j \neq i} S_{ij}) \qquad \text{(Def. IV.8)}$$

and the simplifying notations

$$R_i = R_i^0 + R_i^* \quad \text{and} \quad P_i = P_i^0 + P_i^*, \qquad \text{(Def. IV.9)}$$

Eqn. IV.2 is transformed to

$$n_i + \left(\frac{\mu_i \omega_i}{\Delta_i \nu_i + \Delta_i^*(1-\nu_i)}\right) R_i + \left(\frac{2\,\mu_i}{\delta_i + \delta_i^*}\right) P_i = \mu_i b_i m_i. \qquad \text{(Eqn. IV.3)}$$

The parameters ν_i (Def. IV.3) and ω_i (Def. IV.8) in Eqn. IV.3 are considered as independent, although it is plausible that there may be a simple connection between them. The circumstantial evidence that points in this direction is based on the intuitively apprehended assumption that when the number of bonding MOs for the B^i—B^j bond is governed by one of the atoms, either $S^0_{ij} = b_i s_{ij}$ or $S^0_{ij} = b_j s_{ji}$ (cf. Defs. III.4 and III.5) is satisfied and hence $\omega_i = \nu_i$ or $\omega_i = 1 - \nu_i$. (If both atoms equally govern the situation, Hyps. III.1 and III.2 ensure that $\omega_i = \nu_i = \frac{1}{2}$.)

Eqn. IV.3 is somewhat analogous to expressions arrived at earlier (say, Eqn. II.12); the essential novelty is that Eqn. IV.3 emphasizes the valence situation of each individual B^i. In the standard application of MO descriptions to particular compounds there is implicit an adjustment procedure that is equivalent to an evaluation of an appropriate value for each parameter in Eqn. IV.3. The advantage of Eqn. IV.3 is thus that one is forced explicitly to perform each of these specializations.

An interesting simplified version of Eqn. IV.3 is obtained on introduction of the additional assumptions of Hyps. III.1 and III.2, which give $\nu_i = \omega_i = \frac{1}{2}$, and Hyp. III.5, which implies that $\Delta_i = \delta_i = 2$ and $\Delta^*_i = \delta^*_i = 0$. If, moreover, the non-bonding B^i levels are completely filled, Eqn. IV.3 reduces to

$$n_i + P_i = b_i(2m_i). \qquad \text{(Eqn. IV.4)}$$

This will be recognized as the expression nearest to Eqn. II.2 obtainable on the basis of Def. III.1 .The fact that Q is absent in Eqn. IV.4 compared to Eqn. II.2 emphasizes its insignificance in Eqn. II.2 as well. Note that no assumption concerning electron pairs (Hyp. III.3) has been made in establishing Eqn. IV.4. Moreover, even in cases where Hyps. III.1, III.2, and III.5 are rejected, the bonding situation may comply with Eqn. IV.4 for one or more of the constituents, because of the numerous choices which can be made for the parameters involved (cf. the discussion that follows Eqn. II.12).

A drawback associated with Eqn. IV.3 is that this expression is more concerned with counting orbitals than electrons. In order to improve this situation, a further description of the bonds originating from B^i is introduced through Defs. III.2, III.3, and III.7, thus permitting a reformulation of Eqn. IV.2

$$n_i = \sum_{j \neq i} (b_i v_{ij} + b_j v_{ji}) + b_i v_{ii} + b_i z_i, \qquad \text{(Eqn. IV.5)}$$

where

$$z_i = g_i - v_i \; (= \mu_i u_i) \qquad \text{(Eqn. IV.6)}$$

is the number of non-bonding B^i electrons. If Hyp. III.3 is introduced, Eqn. IV.5 transforms to

$$n_i = b_i(2\theta_i k_i - \theta_{ii}k_{ii} + z_i) \qquad \text{(Eqn. IV.7)}$$

in which $k_i = \sum_j k_{ij}$ and $\theta_i = (\sum_j \theta_{ij}k_{ij})/k_i$. The counting procedure represented by Eqn. IV.7 will be termed the *electron-pair formalism.*

As an alternative, but equivalent expression to Eqn. IV.5, one can take

$$n_i = b_i g_i + \sum_{j \neq i} b_j v_{ji}, \qquad \text{(Eqn. IV.8)}$$

which facilitates the introduction of Hyp. III.6 rather than Hyp. III.3, thus giving

$$n_i = b_i(g_i + v_i - v_{ii}). \qquad \text{(Eqn. IV.9)}$$

The counting procedure represented by Eqn. IV.9 is termed the *neutral-bonded formalism,* and may (like Eqn. IV.7) be regarded as a set of equations supplementary to those presented in Eqn. IV.3 for the generalized (8-N) rule formalism.

To illustrate similarities or differences between electron-pair formalism and neutral-bonded formalism, it is convenient to consider a few apparently typical examples, chosen such that the simplified version Eqn. IV.4 of the general Eqn. IV.3 is supposedly valid for at least one of the compound constituents (say, Sb, Te, or I) of the (isoelectronic) compounds GaSb, ZnTe, and CuI. The bonding situation in the (possibly hypothetical) corresponding diatomic molecules would probably be formulated in terms of such classically predetermined valences (Section II.1) as Ga≡Sb, Zn=Te, and Cu—I. The electron-pair formalism and the neutral-bonded formalism both comply with this picture. However, in the solid state, the three compounds are isostructural with the ZnS-wurtzite and/or -zincblendetype atomic arrangement(s), although, due to the tetrahedral coordination of the atoms, their valence situations are commonly believed to be quite different. The combined application of Eqn. IV.4 (with $P_i = 0$ and $2m_i = 8$) and Eqn. IV.7 (with $\theta_i = 1$, $k_{ij} = 4$, and $k_{ii} = 0$) leads to $z_i = 0$ for Sb, Te, and I. Hence, if the electron-pair formalism is accepted, valence states of Sb(V), Te(VI), and I(VII) are obtained. (This consequence is traditionally not recognized because anionic formulations are adopted for Sb, Te, and I, although clearly contradicted by experimental evidence.) The combined application of Eqn. IV.4 (with $P_i = 0$ and $2m_i = 8$) and Eqn. IV.9 (with $v_{ii} = 0$) gives the valences Sb(III), Te(II), and I(I), and hence Ga(III), Zn(II), and

Cu(I), respectively, in accordance with the requirement of the neutral-bonded formalism.

The characteristic feature that emerges from the combined application of the generalized (8-N) rule and neutral-bonded formalism for the above examples is that the valences deduced do not provide enough electrons for allocation of an electron pair to each of the tetrahedral bonds. This reveals the fundamental distinction between the neutral-bonded and electron-pair formalisms in that the former has no *a priori* requirement of one electron pair per bond. The new approach to chemical bonding consequent on the introduction of quantum-theoretical ideas is shaped by the related concepts of delocalization and exchange of electrons between two or more atoms. These ideas are seen to be preserved in the neutral-bonded formalism which is merely a procedure for counting the collectively behaving B^i—B^j bonding and anti-bonding electrons. It should be emphasized that the neutral-bonded formalism *inter alia* does not take sides in questions concerning the location of valence electrons in space nor their paired status. Hence, the general situation described by the neutral-bonded formalism may well include both the more conventional electron-pair picture and the idea of fractional bonds (through resonating electron pairs, etc.). (Although the resonance concept is fully acceptable, this pictorial visualization will probably be found to have prevented rather than promoted progress.)

The use of the term "neutral-bonded" may be somewhat misleading in that (cf. Def. III.8) the compound constituents cannot be considered effectively neutral unless the bonding situation is assumed to be completely covalent. Clearly, the neutral-bonded formalism (in common with the generalized (8-N) rule and the electron-pair formalism) does not convey information about bonding states. In order to substantiate this statement, it is necessary to show that the formalism is valid also when the atoms possess effective charges. Taking Eqn. III.4 as a convenient starting point, $\lambda'_{ji} > \lambda'_{ij}$ describes a bonding situation in which the B^j constituents initially have a greater attraction for the $(b_i v_{ij} + b_j v_{ji})$ electrons relative to the free atoms than do the B^i constituents. There should accordingly be an increased probability of finding these electrons in the neighbourhood of the B^j atoms, which in turn results in an increased average shielding of the nuclear charge on B^j. At equilibrium, the average electron in the B^i—B^j bonds does not feel any ionic component, but merely the balanced force from the effective nuclear charges on B^i and B^j atoms. The distinction between a completely covalent and a more electrovalent B^i—B^j bond concerns the charge distribution, but not the number of electrons involved. Hence, it is substantiated that the neutral-bonded formalism is indifferent to the appearance of effective charges, which is an external (long-range) effect.

V. Relation to Experimental Data

Sections II to IV have been mainly about various aspects of chemical bonding on the "theoretical" level. Bonding considerations of this kind easily acquire an abstract, speculative character if not related to concrete experimental data. It seems somewhat artificial to extend this paper by a comprehensive discussion of particular examples, which in any case will be the subject of a series of forthcoming papers. However, this article cannot be terminated without a brief survey of the relation between the different bonding parameters and experiments in general. This task can unfortunately only be undertaken on a rather superficial level.

The bonding parameters invoked in Sections II to IV may be roughly divided into five categories with reference to

(i) composition
(ii) structure
(iii) bonded atoms
(iv) bonding character
(v) occupation number.

These categories are listed in order of an apparently increasing degree of experimental inaccessibility.

At first sight, the determination of composition (i.e. the establishment of the empirical chemical formula) seems to be a rather trivial problem within the domain of analytical chemistry. However, if we ask how accurately must the composition be known, it becomes clear that compositional problems may not be trivial after all. It is perhaps sufficient in this context to mention cues like homogeneity/inhomogeneity, purity/impurity, stoichiometry/non-stoichiometry, etc. Of course, the compositional parameters are interwoven with all other bonding parameters and a certain amount of prior knowledge about composition is required for the interpretation of experiments designed to provide information about parameters belonging to categories (ii) to (v). Conversely, these experiments will in turn convey information about composition.

The determination of structure (in the sense of molecular architecture) may in itself present problems of a complexity ranging from the trivially simple to the virtually impossible. The most versatile experiments for determining structure are based on diffraction methods, where fairly general recipes are available for processing the experimental data. This processing in itself (apart from experimental errors) gives rise to both inadequate approximations and illogical feedback loops. Despite these complicating features, the real difficulties are encountered when evaluation of bonding parameters is attempted. It seems appropriate to maintain that only atoms which are separated by comparatively short interatomic distances can be subject to significant bonding interactions; it is therefore natural to distinguish between atoms that are likely to be bonded together and atoms to which the opposite applies.

At present, it is impossible to obtain a direct experimental answer to the important question whether there is or is not bonding between two given atoms. The estimation of bonding is thus confined to deductions from the atomic arrangement and accordingly depends on empirical knowledge. The various schemes which relate interatomic distances to bonding parameters all assume an essentially spherical model for the atoms. The interatomic distances are considered to be additively built up from the radii of the atoms in question and supposedly determined only by bonding parameters. The sum of the radii is, firstly, used to assess whether bonding occurs between given atoms, and secondly, to estimate bond strengths in terms of e.g. bond order. Hopefully we can, without being accused of underestimating the importance of empirical knowledge, say it is fair to attach question marks to the various scales of radii. The interrelation between the three commonly used scales appropriate to ionic, covalent, and metallic bonding has been critically examined by *Slater* (*37*), who concludes that their individual significance is doubtful anyway, because only composite quantities (sums of radii, if the model is acceptable) are susceptible to observation. However, regardless of the significance of the radius concept, it is appropriate to ask whether the number of electrons is the determining factor for bond lengths, or whether factors like number and/or kind of orbitals involved for the atoms in question and other bonded neighbours are of comparable importance. Other factors like the high- or low-spin states for transition, lanthanide, and actinide elements are known to have a profound influence on bond lengths in particular cases.

Accurate determination of absolute electron densities is a natural target for diffraction experiments. Although promising progress has recently been noted, this variant of the traditional diffraction methods is not capable of providing credible information on bond strengths. Apart from the diffraction methods, various spectroscopic techniques are capable of providing bonding information, most generally through exploration of vibrational parameters. In particular and favourable cases, other methods (*e.g.* the different magnetic resonance techniques, ESCA, Mössbauer spectroscopy, etc.) may provide bonding information of equal importance to those mentioned above.

No method capable of giving an experimental determination of the number of electrons participating in bonding is yet available. Even if the problems associated with the determination of absolute electron densities (see above) are solved, one is faced with the problem of distinguishing between electrons of different bonding character, *i.e.* non-bonding core and valence electrons, as well as bonding and anti-bonding electrons. Lacking a general approach to this problem, one is constrained to consult special examples from which more general extrapolations may be possible. As for non-bonding valence electrons, their number may be provided by the combined application of various magnetic investigation techniques. No such approach can be recommended for the separate evaluation of bonding and anti-bonding electrons. Hence, the general situation with respect to bonding character and occupation number is most unsatisfactory.

VI. Summary and Conclusion

This article has been devoted to some aspects of chemical bonding that emerge from the classic key concept of valence. Mathematical formulations have frequently been utilized in order to facilitate precise state ments, so that a brief non-mathematical summary may be called for.

Semi-empirical rules relating the composition, structure, etc. of a given compound to the valence states and features of the electron configurations of the electronegative (anionic) and electropositive (cationic) constituents are commonly used in systematization and prediction. Among these, variants of the (8-N) rule were formulated on the assumption of complete octet configurations around the anionic constituents. Subject to generalizations based on "classic" valence concepts, valence and the number of valence electrons can be treated asymmetrically for the anionic and cationic constituents, respectively. In order to clarify the physical content of the thus derived generalized (8-N) rule, the "classic" valence treatment must be followed up by molecular-orbital considerations. Such an analysis reveals the somewhat arbitrary nature of the common formulation of the rule.

On shifting our attention from the anionic to the cationic constituents, we derive a generalized (8-N) rule also for this part of the molecular system. If the original version is regarded as the normal, the latter may be considered an inverted version of the generalized (8-N) rule. For the counting of electrons, a normal ionic description can formally be used. In connection with the application of the inverted rule, the cationic constituents may, on the other hand, formally be supposed to acquire these electrons (this would require the adoption of an inverted ionic description). The characteristic feature of the generalized (8-N) rule *formalism* is that it provides a counting procedure which leaves the actual bonding states completely unspecified. Closely associated with this, is the question of a general distinction between anionic and cationic constituents, which in turn has led to an artificial classification of bonds in the categories anion–anion, anion–cation, and cation–cation. However, the fundamental question concerns whether there is or is not bonding interaction between two given atoms.

The grouping of three or more components of a given compound into only two categories is in itself an oversimplification, and a more natural classification operates with one class for each kind of atom that can be chemically and structurally distinguished. As a consequence, the asymmetric treatments of valence and number of valence electrons per atom disappear. Within this framework, the generalized (8-N) rule formalism gives rise to a set of equations, one for each kind of atom.

To obtain information supplementary to the generalized (8-N) rule formalism, it is necessary to examine the principles on which the concept of bonding rests. In qualitative descriptions using molecular orbital theory, too much emphasis is usually placed on the empty orbital skeleton, neglecting the fact that electrons, not empty orbitals, are subjected to bonding interactions and that the physical significance of orbital mixing stems from the exchange of electrons. When attention is focused on the electrons, this commonly (and almost automatically) implies the adoption of the electron-pair bond as a fundamental principle. From this principle, a counting procedure termed the *electron-pair formalism*, is derived. However, in view of the large number of macromolecular inorganic compounds that are not in accordance with the electron-pair formalism, this counting procedure is of somewhat limited applicability as a supplement to the generalized (8-N) rule formalism. Moreover, the simultaneous assumption of an electron pair and of complete exchange of electrons within the bond leads paradoxically to highly polar bonds in numerous cases. These, and other unsatisfactory implications arising from the application of the electron-pair idea, enforce its rejection as a fundamental bonding principle, although it may be adequate in special cases as a derived feature.

In cases where the electron-pair formalism leads to unsatisfactory bonding descriptions, and also in a number of cases where this formalism may be adequate, another bonding principle appears to be fulfilled. On the basis of this principle, an additional counting procedure, termed the *neutral-bonded formalism*, is constructed. Although not explicitly recognized as such, the principle on which the neutral-bonded formalism rests is, like the other bonding principles, generalized from quantum-chemical treatments of very simple systems. The simultaneous assumption of the neutral-bonded formalism and complete exchange of electrons within the bonds will produce non-polar bonds, thus constrasting with the electron-pair formalism.

VII. References

1. *Kossel, W.:* Ann. Phys. *49*, 229 (1916).
2. *Lewis, G. N.:* J. Am. Chem. Soc. *38*, 762 (1916).
3. *Langmuir, I.:* J. Am. Chem. Soc. *41*, 868, 1543 (1919).
4. *London, F.:* Z. Phys. *46*, 455 (1928).
5. *Slater, J. C.:* Phys. Rev. *35*, 514 (1930).
6. — Phys. Rev. *37*, 841 (1931).
7. — Phys. Rev. *38*, 1109 (1931).
8. *Pauling, L.:* J. Am. Chem. Soc. *53*, 1367, 3225 (1931).
9. *Hund, F.:* Z. Elektrochem. *34*, 437 (1928).
10. — Z. Phys. *74*, 429 (1932).
11. *Mulliken, R. S.:* Chem. Rev. *9*, 841 (1931).
12. — Phys. Rev. *40*, 55 (1932).
13. — Phys. Rev. *41*, 49, 751 (1932).
14. — Phys. Rev. *43*, 279 (1933).
15. — J. Chem. Phys. *1*, 492 (1933).
16. — J. Chem. Phys. *3*, 375 (1935).
17. *Hume-Rothery, W.:* Phil. Mag. *9*, 65 (1930).
18. *Mooser, E., Pearson, W. B.:* J. Electronics *1*, 629 (1956).
19. — — Progr. Semicond. *5*, 103 (1960).
20. *Hulliger, F., Mooser, E.:* J. Phys. Chem. Solids *24*, 283 (1963).
21. *Pearson, W. B.:* Acta Cryst. *17*, 1 (1964).
22. *Kjekshus, A.:* Acta Chem. Scand. *18*, 2379 (1964).
23. *Hulliger, F., Mooser, E.:* Progr. Solid State Chem. *2*, 330 (1965).
24. *Suchet, J. P.:* Chemical physics of semiconductors. New York: Van Nostrand 1965.
25. *Parthé, E.:* In: *J. Westbrook* (ed.), Intermetallic compounds, p. 180. New York–London–Sydney: Wiley 1967.
26. *Goryunova, N. A., Parthé, E.:* Mater. Sci. Eng. *2*, 1 (1967).
27. *Hulliger, F.:* Struct. Bonding *4*, 83 (1968).
28. *Parthé, E.:* Acta Cryst. *B 29*, 2808 (1973).
29. *Löwdin, P.-O.:* J. Phys. Chem. *61*, 55 (1957).
30. *Wilson, A. H.:* Proc. Roy. Soc. (London) *A 133*, 458 (1931).
31. *Goodman, C. H. L.:* J. Phys. Chem. Solids *6*, 305 (1958).
32. *Bushman, E.:* Z. Anorg. Allgem. Chem. *313*, 90 (1961).
33. *Goodman, C. H. L.:* Nature *187*, 590 (1960).
34. *Mooser, E., Pearson, W. B.:* Nature *190*, 406 (1961).
35. *Goodman, C. H. L.:* Nature *192*, 355 (1961).
36. *Mooser, E., Pearson, W. B.:* Nature *192*, 355 (1961).
37. *Slater, J. C.:* J. Chem. Phys. *41*, 3199 (1964).

Received February 4, 1974

Geometrical Considerations on the Marcasite Type Structure

A. Kjekshus and **T. Rakke**

Department of Chemistry, University of Oslo, Blindern, Oslo 3, Norway

Table of Contents

I. Introduction

The structural and bonding relationships between the FeS_2—*p* (*p* = pyrite) type on the one hand, and the three classes A, A/B (including FeAsS-arsenopyrite; binary prototype $CoSb_2$), and B of the FeS_2—*m* (*m* = marcasite) type on the other, have previously been discussed by one of the present authors (*1*, *2*). The novelty in treatment was summarized in two somewhat interrelated models termed the pair-reorientation model (*1*) concerned with the geometrical relationship between the structure types FeS_2—*p* and FeS_2—*m*, and the expansion model (*2*) relating the different classes of the FeS_2—*m* type to the configuration (d^j) of localized electrons on the transition-metal (*T*) atoms. The latter model has recently been subjected to justified criticism by *Goodenough* (*3*), although he apparently accepts the former model. A further refinement of Goodenough's bonding considerations is called for, and this will be the subject of a separate paper (*4*). After being reminded by *Goodenough* of defects in the expansions model, we felt the need for a critical reexamination of the pair-reorientation model. These analyses have brought out a number of interesting points of view, which are presented here.

II. Geometrical Relations between the Types FeS_2–*p* and FeS_2–*m*

Any attempt to relate the structure types FeS_2—*p* and FeS_2—*m* must utilize their common characteristics: nominal composition TX_2 ($X =$ pnigogen or chalcogen), octahedral coordination of T, tetrahedral coordination of X, and X—X pairs. As already pointed out (*1*), regular coordination polyhedra are incompatible with these characteristics, and the actual shapes of the coordination polyhedra may be influenced by such factors as competing symmetry requirements for T and X, details of the packing arrangements, etc. If we neglect some of these factors, significant features of the two structure types can be summarized in the expressions for the bond distances:

FeS_2—*p* [cubic, *Pa*3, cf. (*5*)] Def. 1

$$T\text{—}X:\ d_p = a_p\left[3x_p^2 - 2x_p + \frac{1}{2}\right]^{\frac{1}{2}}$$

$$X\text{—}X:\ l_p = 2\sqrt{3}\,a_p\left(\frac{1}{2} - x_p\right)$$

FeS_2—*m* [orthorhombic, *Pnnm*, cf. (*6*)] Def. 2

$$T\text{—}X:\ d_1 = \left[a^2x^2 + b^2y^2\right]^{\frac{1}{2}}$$

$$d_2 = \left[a^2\left(\frac{1}{2} - x\right)^2 + b^2\left(\frac{1}{2} - y\right)^2 + c^2\left(\frac{1}{2}\right)^2\right]^{\frac{1}{2}}$$

$$X\text{—}X:\ l = 2\left[a^2x^2 + b^2\left(\frac{1}{2} - y\right)^2\right]^{\frac{1}{2}}$$

The notable features of the pair-reorientation model (*1*) can be summarized as follows:

(i) The recognition of the virtually identical atomic arrangements in (001) and ($\bar{1}01$) of the FeS_2—*p* and FeS_2—*m* type structures, respectively.

(ii) An imagined phase transformation from the FeS_2—*p* to the FeS_2—*m* type structure involves *inter alia* a reorientation of alternating X—X pairs along (say) directions parallel to [011] in the former.

(iii) The arrangement according to (ii) would result in a tetragonally shaped FeS_2—*m* type cell defined by the transformation matrix $\left\{0, \frac{1}{2}, -\frac{1}{2} / 0, -\frac{1}{2}, -\frac{1}{2} / -1, 0, 0\right\}$.

(iv) The geometrical shape of the X—X pair approximately resembles a short sausage, *i.e.* a cylinder of diameter and length $2l_p$, terminated by two hemispheres of the same diameter, so that the total length of the thus defined sausoid is $2l_p$.

(v) The further transformation of the imaginary, tetragonally shaped FeS_2—*m* type cell to one of orthorhombic symmetry is a "consequence" of the sausoidal shape of the X—X pair.

Reexamination of the pair-reorientation model is most conveniently carried out on the basis of mathematical formulations of all explicit and implicit assumptions.

The postulated correspondence in bond lengths for the two structure types leads to the four equations

$$l_p = l \qquad \text{Hyp. 1}$$

$$d_p = d_1 \qquad \text{Hyp. 2}$$

$$d_p = d_2 \qquad \text{Hyp. 3}$$

$$d_1 = d_2. \qquad \text{Hyp. 4}$$

The assumption of Hyp. 4 only, gives rise to the relation

$$\left(\frac{a}{b}\right)^2 x + y = \frac{1}{4}\left[1 + \left(\frac{a}{b}\right)^2 + \left(\frac{c}{b}\right)^2\right], \tag{1}$$

which is [cf. (7)] approximately satisfied in known marcasite series. Hyps. 1 to 3 are composite multiparameter relations from which no further deductions can be made without additional assumptions. In Ref. (*1*), the supplementary conditions:

$$a_p = b \qquad \text{Hyp. 5}$$

$$a^2 + c^2 = b^2 \qquad \text{Hyp. 6}$$

were formulated explicitly, and implicitly:

$$x_p = y. \qquad \text{Hyp. 7}$$

Hyps. 1 to 7 are not independent in that Hyp. 4 trivially follows from Hyps. 2 and 3, and Hyp. 7 can be deduced by combining Hyps. 1 to 3, 5, and 6. Hyp. 5 is indispensable in this connection since it facilitates a simple conversion between the two types of structure. For class B of the FeS_2—*m* type, this hypothesis is a reasonable approximation and the a_p/b ratio is consistently observed to vary between 0.987 ($CuSe_2$) and 1.004 ($FeTe_2$). Hyp. 5 on the other hand expresses an entirely hypothetical relation for class A, where no FeS_2—*p* modifications have hitherto been observed, and the postulate must therefore be regarded with suspicion for this class. The experimental data for both classes of compounds with

FeS_2—*m* type structure confirm that Hyp. 6 is approximately satisfied, the ratio $(a^2+c^2)/b^2$ varying between 0.995 ($CrSb_2$) and 1.035 ($FeSb_2$) in class A, and between 1.023 ($CuSe_2$) and 1.093 ($CoTe_2$) in class B. The situation concerning Hyp. 7 is similar to that for Hyp. 5 but, since FeS_2 is the only compound for which accurate positional parameters are available for both modifications [$x_p=0.3840(5)$ and $y=0.37820(5)$, cf. (*5*, *6*)], further experimental tests of the degree of validity of this postulate are called for. Note that the less accurate parameters for NaO_2 satisfy (*8*) the relation exactly ($x_p=y=0.43$).

Based on simple pictorial arguments, the pair-reorientation model introduced the expression

$$c = a_p/\sqrt{2} - l_p/2 \qquad \text{Hyp. 8}$$

for the shortening of the *c* axis of the relaxed FeS_2—*m* cell. Finally, the so-called ideal value

$$x_p = 3/8 \qquad \text{Hyp. 9}$$

was taken as a representative average for the positional parameter of the FeS_2—*p* type atomic arrangement.

By choosing different combinations of Hyps. 1 to 9, various "relations" between the FeS_2—*p* and FeS_2—*m* type atomic arrangements may be deduced, but only those relevant to the judgement of the pair-reorientation model are presented here. The combination of Hyps. 1, 5, and 7 gives an expression

$$x = \sqrt{2}(b/a)\left(\frac{1}{2}-x_p\right), \qquad (2)$$

which can be used to relate bonding *T*—*X* distances in the two structure types. Hyp. 2 is now a derived feature and

$$d_2^2 = d_p^2 + (1/4)(a^2+b^2+c^2-2\sqrt{2}ab) - b(b-\sqrt{2}a)x_p \qquad (3)$$

follows as an additional consequence. The latter expression can be simplified to

$$d_2^2 = d_p^2 + b(b-\sqrt{2}a)\left(\frac{1}{2}-x_p\right) \qquad (4)$$

on the introduction of Hyp. 6.

Alternatively, the combination of Hyps 4 to 6 gives

$$x = (b/a)^2 \left(\frac{1}{2} - x_p\right), \tag{5}$$

which in combination with Eq. (2) (using in total Hyps. 1 and 4 to 7) gives

$$b = \sqrt{2}a; \tag{6}$$

subsequently, Hyp. 3 follows from Eq. (4), and Hyp. 6 can be further simplified to

$$a = c = b/\sqrt{2}. \tag{7}$$

Analysis shows that Hyps. 1 to 7 are mutually consistent, whereas Eq. (7) is clearly contradictory to Hyp. 8. The incompatibility of Hyp. 8 with Hyps. 1 to 7, indicates that a logical inconsistency must have been present in the original derivation of the pair-reorientation model. In fact, the cell with axes defined by Eq. (7) is identical with the tetragonally shaped FeS_2—*m* type cell, which was regarded as the imaginary intermediate in the hypothetical transformation route described by the pair-reorientation model (see (*1*) and above). Furthermore, on combining Hyps. 1 to 7 with Hyp. 9, it is seen that

$$x = 1/4 \text{ and } l = c, \tag{8}$$

which implies that the simple pairing of the X atoms cannot be retained. There is thus another inconsistency in the original derivation of the pair-reorientation model.

The above treatment shows that one or more of Hyps. 1 to 9 must be removed. Axial ratios ($c/a \approx 0.56$ and $c/b \approx 0.49$) corresponding to those observed in class A can be reached through the mere application of Hyps. 1, 5, 6, 8, and 9 [cf. (*1*)]. However, this result is obtained at the expense of equal T—X bond distances, implying that Hyps. 2 to 4 *must* be relinquished. (As is evident from the above, the requirements of Hyps. 2 to 4 are intimately coupled to the tetragonal shape of the FeS_2—*m* type cell.) An immediate consequence of this is that, contrary to the original inference (*1*), the relationship between the FeS_2—*p* and FeS_2—*m* type structures cannot be governed by the difference in X—X packing alone.

It is an experimental fact that structural transformations between the FeS_2—*m* and FeS_2—*p* type atomic arrangements are exclusively observed for compounds belonging to class B (including NaO_2) of the former, whereas no compound in class A has yet been found to have a corresponding FeS_2—*p* type modification. Hence, the original version of the pair-reorientation model is in the paradoxical situation that it relates hypothetical FeS_2—p modifications to class A, while well-known transformations between FeS_2—p and class B cannot be simulated by the model. This naturally evokes the question whether the model can be modified to cope with the latter, experimentally founded situation.

In order to answer this question, it is convenient to choose as a starting point the tetragonally shaped FeS_2—m type cell, as defined in (iii) on p. 86. An orthorhombic deformation of this cell in line with the experimental data would require an increase of a and/or a decrease of c according to

$$a = b/\sqrt{2} + \delta_a \tag{9}$$

$$c = b/\sqrt{2} - \delta_c . \tag{10}$$

The construction of Eqs. (9) and (10) assumes implicitly that Hyp. 5 is a valid approximation, and if Hyp. 6 is also accepted, it follows that

$$\sqrt{2}\,(\delta_a - \delta_c)\,b + \delta_a^2 + \delta_c^2 = 0, \tag{11}$$

which requires $\delta_a < \delta_c$. The empirical analogues of Eqs. (9) and (10) for class B (averaged over all members) are

$$a \approx b/\sqrt{2} + 0.125\,b \tag{12}$$

$$c \approx b/\sqrt{2} - 0.088\,b\,; \tag{13}$$

thus, it is immediately evident that Eq. (11) cannot be fulfilled. The cause of the trouble for class B is clearly the acceptance of Hyps. 5 and 6 as appropriate approximations. In reality, the situation is even worse for class A, where Hyp. 5 represents a purely imaginary relation rather than an approximation.

It follows from the analyses presented here that the previously advocated inference (*7*) that class A is to be considered as the "normal" class of the FeS_2—m type atomic arrangement (resulting from the type of packing of the X—X pairs) may no longer be valid. (It should be emphasized that the quantitative shortcomings of the pair-reorientation model cannot be remedied by reformulating Hyp. 8, even if this represented an improved description of the electron distribution around an X—X pair.) Although the quantitative content of the pair-reorientation model must be rejected, its qualitative content as a hypothetical transformation route relating the atomic arrangements of the FeS_2—p and FeS_2—m type structures is still of some value.

III. Quasi-hexagonal Close-packing of $X(X_2)$ in the FeS_2–*m* Type

One aspect that remains to be explored is the extent to which the packing of X atoms alone governs the structural characteristics of the FeS_2—*m* type. As variously pointed out in the literature [*e.g.* (*9, 10*)], the location of the X atoms in the FeS_2—*m* type structure, in common with the TiO_2—*r* (*r* = rutile) type, bears some resemblance to hexagonal close-packing.

As a first approximation, let us consider the X atoms as rigid spheres arranged in an ideal hexagonal close-packed (*hcp*) sublattice. On transforming the usual hexagonal unit cell ($\boldsymbol{a}_h, \boldsymbol{b}_h, \boldsymbol{c}_h$) to an ortho-hexagonal cell, defined by $\boldsymbol{a}_0 = \boldsymbol{c}_h$, $\boldsymbol{b}_0 = \boldsymbol{b}_h - \boldsymbol{a}_h$, and $\boldsymbol{c}_0 = \boldsymbol{a}_h + \boldsymbol{b}_h$, and introducing T on the octahedral sites in an ordered manner, as indicated in Fig. 1a,

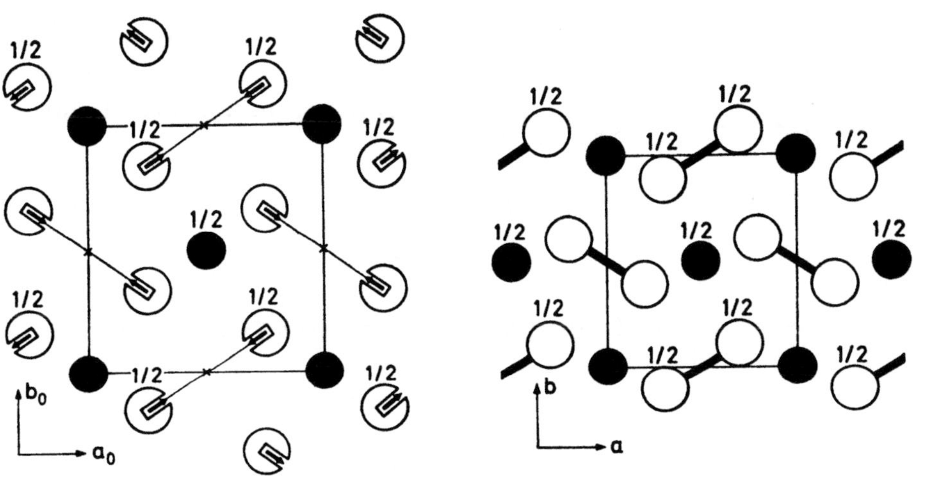

Fig. 1. Atomic arrangement of X (open circles) and T (filled circles) in projection for: (a) hexagonal close-packing of X with T occupying half the octahedral holes (positions of the other half being indicated by crosses), and (b) the FeS_2—*m* type structure, where the X—X pairs are emphasized by connecting bars

an orthorhombic FeS_2—*m* (or TiO_2—*r*) like cell is obtained. For this atomic arrangement (referred to as the FeS_2—*m*, *hcp* X construction) axial proportions

$$c_0/a_0 = \sqrt{3/8} \approx 0.612 \tag{14}$$

$$c_0/b_0 = \sqrt{1/3} \approx 0.577 \tag{15}$$

that fall between those for class A and class B are derived when the size of T is neglected.

To allow for the formation of X—X pairs, the requirement for rigid spheres must be partly relinquished, each sphere being considered to overlap with one , and only one neighbouring sphere in a pattern prescribed by the FeS_2—*m* type structure (Fig. 1b). The shortening of one X—X distance per X compared with the FeS_2—*m*, *hcp* X construction (Fig. 1a) may be supposed to induce a corresponding shortening in a_0 and b_0. However, c_0 cannot become shorter, since the X—X pair lacks a component along [001] where the X atoms are close-packed already in the FeS_2—*m*, *hcp* X arrangement, consequently

$$c = c_0. \tag{16}$$

As a working hypothesis, it is natural to assume that the octahedral holes in the original *hcp* X sublattice do not change in size, *i.e.* that the bonding T—X interatomic distances are invariant during the imagined transformation from the FeS_2—*m*, *hcp* X to the X–X pair modified (hereafter referred to as FeS_2—*m*, *hcp* X_2) arrangement. The constraints imposed on Def. 2 are accordingly

$$d_1 = c_0/\sqrt{2} \tag{17}$$

$$d_2 = c_0/\sqrt{2}. \tag{18}$$

The application of Eqs. (16) to (18) and the introduction of l from the original Def. 2 give

$$x = \frac{1}{4}\left(1 - \frac{c_0^2 - l^2}{a^2}\right) \tag{19}$$

$$y = \frac{1}{4}\left(1 + \frac{2c_0^2 - l^2}{b^2}\right) \tag{20}$$

for the positional parameters of X in the FeS_2—*m*, *hcp* X_2 arrangement.

In a close-packed X sublattice there are twelve nearest and equal X—X distances. The alteration introduced on turning to FeS_2—*m* geometry splits these distances into five, giving in addition to l (Def. 2):

$$l_1 = c \qquad \text{Def. 2}'$$

$$l_2 = 2[(1/2-x)^2a^2 + (1/2-y)^2b^2]^{\frac{1}{2}}$$

$$l_3 = [4(1/4-x)^2a^2 + (b/2)^2 + (c/2)^2]^{\frac{1}{2}}$$

$$l_4 = [(a/2)^2 + 4(1/4-y)^2b^2 + (c/2)^2]^{\frac{1}{2}}.$$

However, in the FeS_2—*m*, *hcp* X_2 arrangement, $l_1 = c_0$ from Eq. 16, and combination of Eqs. (16) and (18) reveals that

$$l_2 = c_0 \tag{21}$$

is automatically fulfilled. Moreover, it follows from Eqs. (17) and (18) that

$$l_3^2 + l_4^2 = 2c_0^2 \tag{22}$$

and, since c_0 is the shortest possible non-bonding X—X distance in FeS_2—*m*, *hcp* X_2,

$$l_3 = l_4 = c_0 . \tag{23}$$

Hence, all the contacts l_1 to l_4 (Def. 2′) still correspond to close-packing in the latter arrangement.

The substitution of Eqs. (19) and (20) into Eq. (23) and the use of Eq. (16) give

$$a^2(3c_0^2 - b^2) = (c_0^2 - l^2)^2 , \tag{24}$$

which can be used to eliminate a or b from Eq. (17). In terms of the variable $\lambda = l/c_0$, the final results are

$$c/a = [3/(3 + \lambda^2 + 2\lambda\sqrt{2(3 - \lambda^2)})]^{\frac{1}{2}} \tag{25}$$

$$c/b = [3/(6 - \lambda^2 + 2\lambda\sqrt{2(3 - \lambda^2)})]^{\frac{1}{2}} \tag{26}$$

the functional dependences of which are presented in Fig. 2. The diagram demonstrates quite clearly that the observed axial ratios for class A are unobtainable within the defined range of λ $(0 < \lambda < 1)$. The axial ratios for class B, on the other hand, fall within the permissible range of λ. Nevertheless, there are fundamental discrepancies for this class too, in that the experimentally determined λ's are found at much higher values than those prescribed by the observed axial ratios according to the FeS_2—*m*, *hcp* X_2 construction [Eqs. (25) and (26)]. By inspection of Fig. 2, NaO_2 is seen to be the only compound that roughly fits into the FeS_2—*m*, *hcp* X_2 scheme. [In this connection it is of some interest to note that, as $\lambda \to 0$, Eqs. (25) and (26) show that $c/a \to 1$ and $c/b \to 1/\sqrt{2}$. On approaching this limit, a more and more NaCl like arrangement is gradually obtained.]

From the above considerations, it should be clear that packing models based on *hcp* X or *hcp* X_2 alone cannot provide appropriate descriptions for compounds with the FeS_2—*m* type structure. Before we reach a

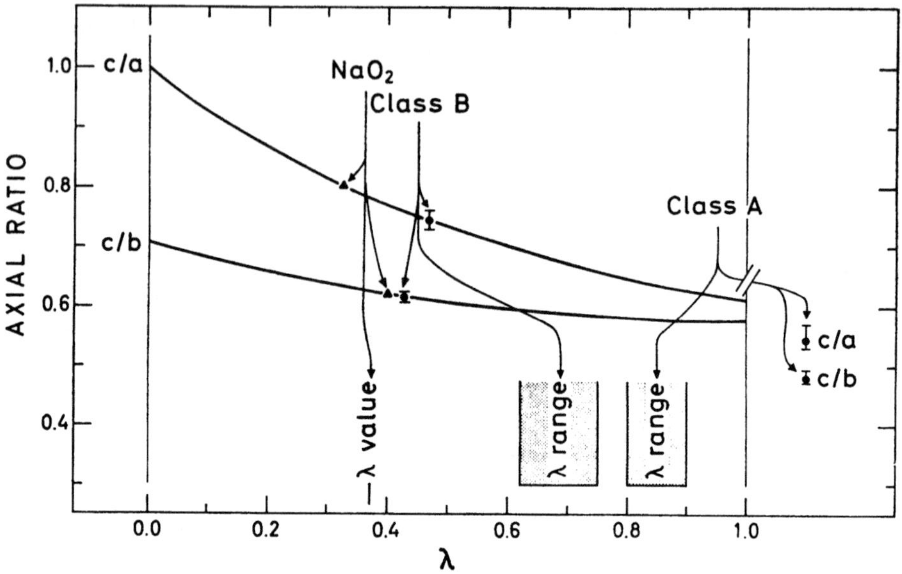

Fig. 2. Axial ratios *versus* λ [Eqs. (25) and (26)] for the FeS_2—*m*, *hcp* X_2 model. Experimentally determined quantities for NaO_2 and average values and scatter for classes A and B are indicated

conclusion on attempts to characterize the FeS_2—*m* classes by X or X_2 packing features, the experimental facts must be briefly consulted. Hence, in Fig. 3, series of TX_2 compounds with constant T are plotted against the principal quantum number of X. The trends within the various series show a pronounced similarity, regardless of whether members of the classes A, A/B (here represented by compounds with the $CoSb_2$ type structure; cf. Introduction), or B are considered. This fact suggests very strongly that, if one of the classes (say, A) is governed solely by the packing features of X or X_2, much the same explanation must apply to the others.

The natural way to seek improvement on the plain FeS_2—*m*, *hcp* X_2 model would be to take into account the T atoms in the form of rigid spheres whose size exceeds those of the original octahedral cavities. (The clearly less realistic FeS_2—*m*, *hcp* X model need not be considered separatly in this connection, but it should be noted that all conclusions for this model would be virtually identical with those drawn for FeS_2—*m*, *hcp* X_2.) However, if the only modifying influence of T was due to its attributed spherical shape, this would produce a uniform overall expansion of all cell dimensions without appreciable alterations in the axial ratios. The thus modified atomic arrangement would fit the experimental

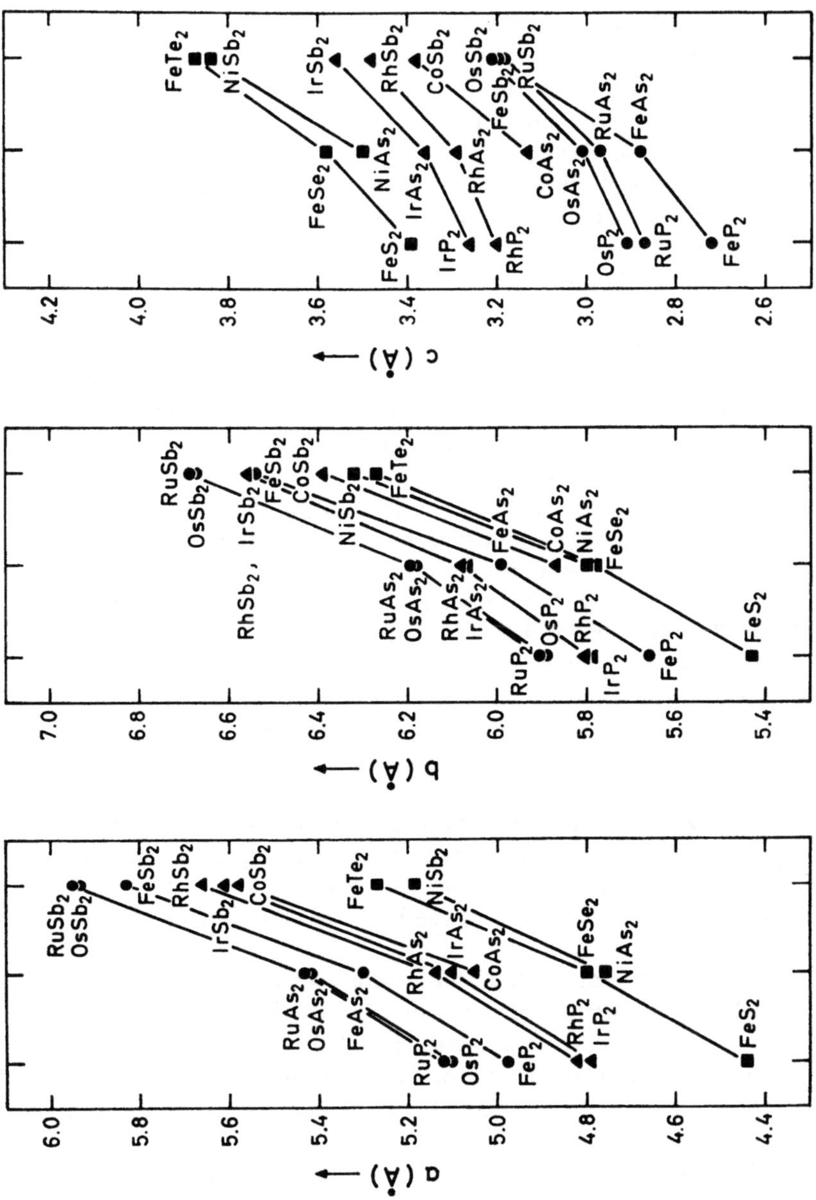

Fig. 3. Cell dimensions for compounds of the structure types FeS_2—*m*, class A (●), $CoSb_2$ (▲), and FeS_2—*m*, class B (■) *versus* principal quantum number (*n*) of *X*. [For the $CoSb_2$ type, axes refer to the pseudo-marcasite variant of the $CoSb_2$ type cell; cf. (*7*)]

facts for class A no better than the plain FeS_2—*m*, *hcp* X_2 model while, for class B, the consequences would be even worse, since the axial ratios would be invariant whereas c_0 (in $\lambda = l/c_0$) would have to be increased to comply with the experimental values for *c*.

Further modification of the FeS_2—*m*, *hcp* X_2 model, *e.g.* by assuming a non-spherical shape for *T*, seems unable to rescue the situation for class B. Class A, on the other hand, is not *a priori* doomed and imagined remedies might be:

(i) The *c* axis is shorter than the close-packing distance c_0. This cannot be due to the shape and size of *T* alone and should consequently be considered as circumstantial evidence for additional *X*—*X* (and possibly *T*—*T*) bonding along [001]. Although *c* is relatively short for compounds of class A, it appears to be too long to reflect any significant degree of direct *X*—*X* and/or *T*—*T* bonding interaction [cf. (1)].

(ii) The *a* and *b* axes are increased relative to *c* as a consequence of of a non-spherical shape of *T*. Thus, the possibility of close-packed *X* atoms would at most be those along [001]. At first, the implication of close-packing along [001] only, seems promising due to the relatively short *c*. As shown in Fig. 4, the appreciable variation of *c* within series with constant *T* shows similar and almost parallel behaviour. This also applies to the data for compounds with the $CoSb_2$ type structure, which have been included for the purpose of comparison. [The deviation from the general trend of *c* for $FeSb_2$ is of no significance in this connection, although it is interesting to note that the unique character of this compound is also reflected in other properties; cf. *e.g.* (*11*).] Provided that the variations of *a* and *b* within one of these series (Fig. 4) are considered significant, the corresponding variation of *c* must be interpreted similarly. This lends further support to the earlier proposal (see above) that class A be rejected as the "normal" class of the FeS_2—*m* type atomic arrangement.

Hence, in order to establish a connection between the FeS_2—*m*, *hcp* X_2 model and the structural facts for class A, the *T* atoms would have to take an almost non-constrained shape. This takes us so far away from the plain *hcp* X_2 model that it seems artificial to pursue the line any further; attempts to modify *this* model with respect to the shape and size of X_2 lead to a similar conclusion.

One further lesson can also be drawn from the entertainment with the FeS_2—*m*, *hcp* X_2 model. Compounds with the FeS_2—*m* type structure (as well as other types with smaller or larger internally connected aggregates of *X* atoms) are normally believed to exhibit an internal measure for the size of *X* (*i.e.* radius in the spherical case), which in turn would permit derivation of the corresponding quantity for *T*. The procedure requires that the radius of *X* with respect to *X* equals that

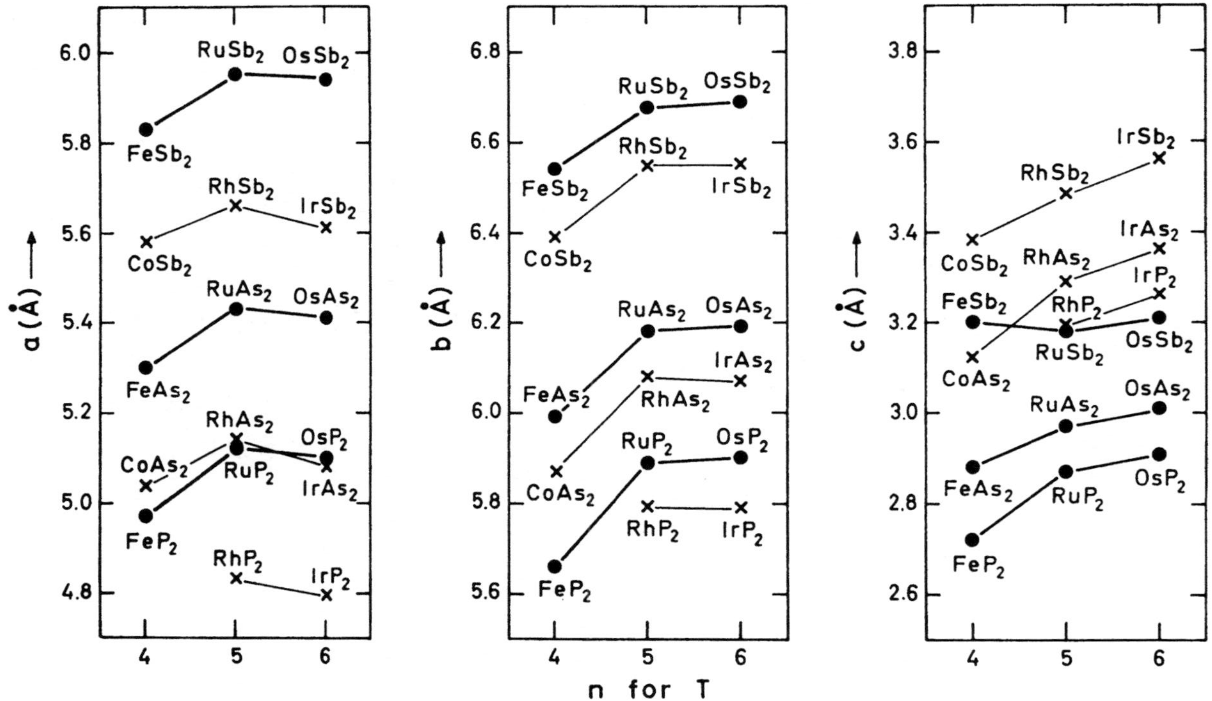

Fig. 4. Dependence of cell dimensions for compounds of the structure types FeS_2—*m*, class A (●) and $CoSb_2$ (×) on principal quantum number (*n*) for *T*

with respect to T, and this represents an unsupported assumption. According to the FeS_2—*m*, *hcp* X_2 model, this assumption can at best represent a relatively coarse approximation for compounds of FeS_2—*m* type structure, a conclusion that is indeed confirmed by analyses of the experimental data.

IV. Relations between the Structure Types FeS_2-*p*, *random* FeS_2-*p*, FeS_2-*m*, CaC_2, and NaCN

In discussions of structural characteristics, compounds with the structure types FeS_2—*p*, FeS_2—*m*, and $CoSb_2$ have been considered as members of an exclusive club in which strangers are permitted only reluctantly. [This statement does not cover the closely related, ordered ternary types CoAsS-cobaltite and NiSbS-ullmanite, which differ from the FeS_2—*p* type mainly in the distribution of X and X' on the X sites of the latter (*12*).] The reasons for this are to be found in their common characteristics (cf. p. 86). The discussions on this subject have hitherto mainly concerned transition metal representatives, among which the bonding characteristics vary surprisingly little. Because of this close relation in bonding, it is likely that inferences about the factors that govern the occurrence of the various structure types may be misleading, and it is therefore of utmost importance that the exclusive requirement for membership in the club be broken. In this generalization, the composition MX_2 and the X—X pair are retained (this permits M to be any metal and, under extreme conditions, *e.g.* high pressure, even this limitation can be inappropriate, and X to be a non-metallic element accepting combinations X, X'), whereas the requirements for octahedral or tetrahedral coordinations can be relinquished.

For the present purpose it is not convenient to go further than to require that the location of M and the center of gravity of the X—X pairs be basically similar and that the orientation of these pairs be describable in a simple way. With these constraints, the considerations appear to be limited to the structure types: FeS_2—*p*, *random* FeS_2—*p*, FeS_2—*m*, CaC_2 [*i.e.* CaC_2(I), cf. *e.g.* (*13*)], and NaCN [*i.e.* the low-temperature modification of NaCN (*14*)]. The close structural resemblance between these structure types is clearly demonstrated in Fig. 5, which also presents hitherto known examples of compounds in which a transformation between the structure types can be induced by varying the external conditions (*e.g.* temperature and/or pressure).

Before we proceed any further, it is appropriate to recapitulate a little on the occurrence of the structure types in question.

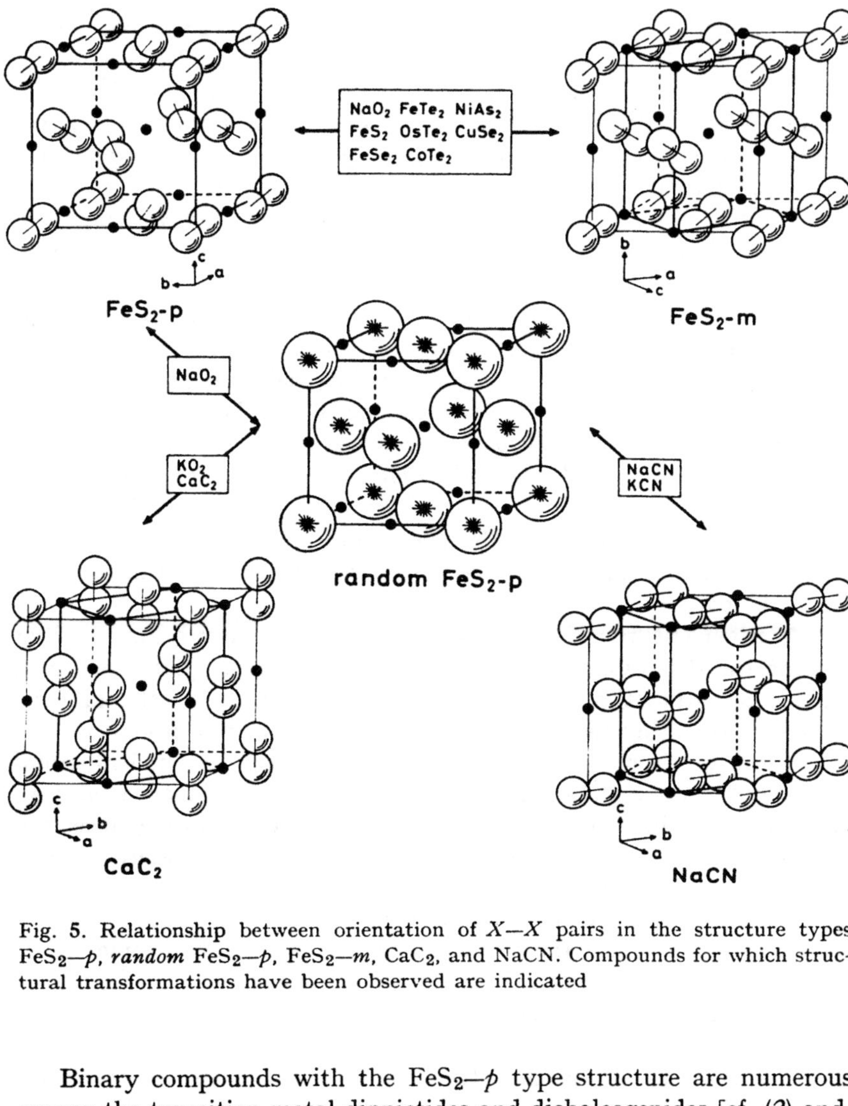

Fig. 5. Relationship between orientation of $X—X$ pairs in the structure types FeS_2—*p*, *random* FeS_2—*p*, FeS_2—*m*, CaC_2, and NaCN. Compounds for which structural transformations have been observed are indicated

Binary compounds with the FeS_2—*p* type structure are numerous among the transition metal dipnictides and dichalcogenides [cf. (*2*) and, include additionally ZnO_2 (*15*) and CdO_2 (*16, 17*)], but other examples like NaO_2 (*8*), MgO_2 (*18*), $MgTe_2$ (*19*), SiP_2 (*20*), and $SiAs_2$ (*20*) are known. Binary compounds with the FeS_2—*m* type structure are even more within the domain of transition metal dipnictides and dichalcogenides [cf. (*2*)] in that the only expection is NaO_2. The latter compound is, moreover, the only compound reported (*8*) to possess all three structure types, FeS_2—*m*, FeS_2—*p*, and *random* FeS_2—*p* (their stability being given in order of increasing temperature).

In addition to a number of carbides [cf. *e.g* (*21—23*)], some oxides [KO_2, RbO_2, CsO_2, CaO_2, SrO_2, and BaO_2; cf. *e.g.* (*24*)] possess the CaC_2 type structure at room temperature. Of these, only KO_2 (*25*) and CaC_2 (*26*) have definitely been observed to transform to the *random* FeS_2—*p* type at elevated temperatures, but there are indications that such transformations (*21*) also occur with SrC_2 and BaC_2.

The NaCN type has hitherto been represented by only two compounds [NaCN (*14*) and KCN (*27*)] stable below room temperature, both of which take on the *random* FeS_2—*p* type structure at room temperature.

The FeS_2—*p* structure can be imagined as built up from a face-centered cubic *M* sublattice, wrapped up in *X—X* pairs oriented in an ordered manner parallel to the four body diagonals of the cube, and with the center of gravity of the pairs occupying the octahedral sites of the *M* sublattice. In the FeS_2—*m* type, on the other hand, the pairs are oriented parallel to only two of the body diagonals of the FeS_2—*p* like cell. Orientation of all pairs parallel to either [001] or [$1\bar{1}0$] of the FeS_2—*p* like cell gives the CaC_2 or NaCN type, respectively. The *random* FeS_2—*p* type (NaCl like arrangement) may be regarded as any of the above types with the pairs oriented at random.

As is evident from Fig. 5, *random* FeS_2—*p* occupies a central position among the structure types in question. This suggests that a relatively simple semi-quantitative explanation for the cause of the different modes of orienting the pairs, and the consequent implications for *e.g.* axial ratios, should be within reach. However, the understanding process must proceed through a number of steps; this involves *inter alia* that decisions have to be taken on some of the following questions:

(i) What geometrical shape and size is most appropriately ascribed to the *X—X* pair?
(ii) What geometrical shape and size applies to *M*?
(iii) To what extent is the structural choice governed by geometry, *i.e.* the mode and efficiency in packing of X_2 and *M*?
(iv) To what extent is the structural choice governed by covalency effects?
(v) To what extent is the structural choice influenced by *effective* atomic charges?

The questions have been listed according to the conventional apprehension of their mutual importance as structure determining factors. The first three questions have the special quality of inviting us to discuss structural features from a purely geometrical point of view. The preceding sections of this paper can perhaps be considered as a rather voluminous treatment of questions (i) to (iii) relative to the FeS_2—*m* (and to some extent FeS_2—*p*) type structure(s). Questions (i) to (iii) have also briefly been taken up for the structure types CaC_2 (*22*) and NaCN (*27*).

A common feature of the latter models is that ions (*e.g.* Ca^{2+}, C_2^{2-}; Na^+, CN^-) are invoked as structural entities. [One of these models (*22*) contains an inherent error in the packing mode of the C_2^{2-} ions, even when viewed in isolation within the subgroup CaC_2, SrC_2, and BaC_2 of the structure type, whereas the other (*27*) is mutually consistent as a model for NaCN and KCN.] However, the relative changes in size and shape between the components in these compounds cannot account for the structural differences. Neither are the distinctions in (formal) ionic descriptions able to provide an explanation, as is nicely demonstrated by the fact that *e.g.* KO_2 can assume the CaC_2 type structure. For the time being it appears fruitful to pursue the working hypothesis that effective, rather than formal ionic charges are a major structure determining factor, *i.e.* to switch the priority invoked in the numbering of the questions and attack first question (v) or the closely related question (iv). In line with this, it is natural to suggest that efforts should be made to find a suitable reference substance.

The fact that NaO_2 possesses three of the structure types under consideration suggests its tentative applicability as an appropriate reference substance. The FeS_2—*m* type modification of NaO_2 has hitherto not found its proper place among isostructural compounds. This neglect of NaO_2 may perhaps be attributed to its outstanding character as a non-transition metal compound, its commonly assumed ionic (Na^+, O_2^-) bonding situation, and/or its relatively large axial ratios ($c/a = 0.80$ and $c/b = 0.62$). However, the axial ratios and normalized axes are, as is evident from Fig. 6, not even approximately constant for compounds within each of the classes A, A/B, and B, but show on the contrary marked and largely similar dependences on the principal quantum number of X. The (say) c/a ratio increases in the series $FeTe_2$—$FeSe_2$—FeS_2 and a hypothetical, isostructural oxide FeO_2 might attain a c/a value close to that for NaO_2. Hence, the FeS_2—*m* type modification of NaO_2 can most naturally be said to belong to class B.

Other arguments that favour NaO_2 as an appropriate reference substance are the presumed more clearcut situation with respect to charge distribution and separability of the X—X pairs in this compound as compared with the transition metal representatives with FeS_2—*p* and FeS_2—*m* type structures. Moreover, *electrons* of *d* and/or *f* character almost certainly do not play a major role in determining the bonding properties of NaO_2. In these respects, the FeS_2—*m* type modification of NaO_2 may be considered to represent the most "normal" FeS_2—*m* type atomic arrangement.

Let us close by expressing the optimistic view that the problems outlined can be solved, and that this paper may perhaps have given a lead in how to reach this goal.

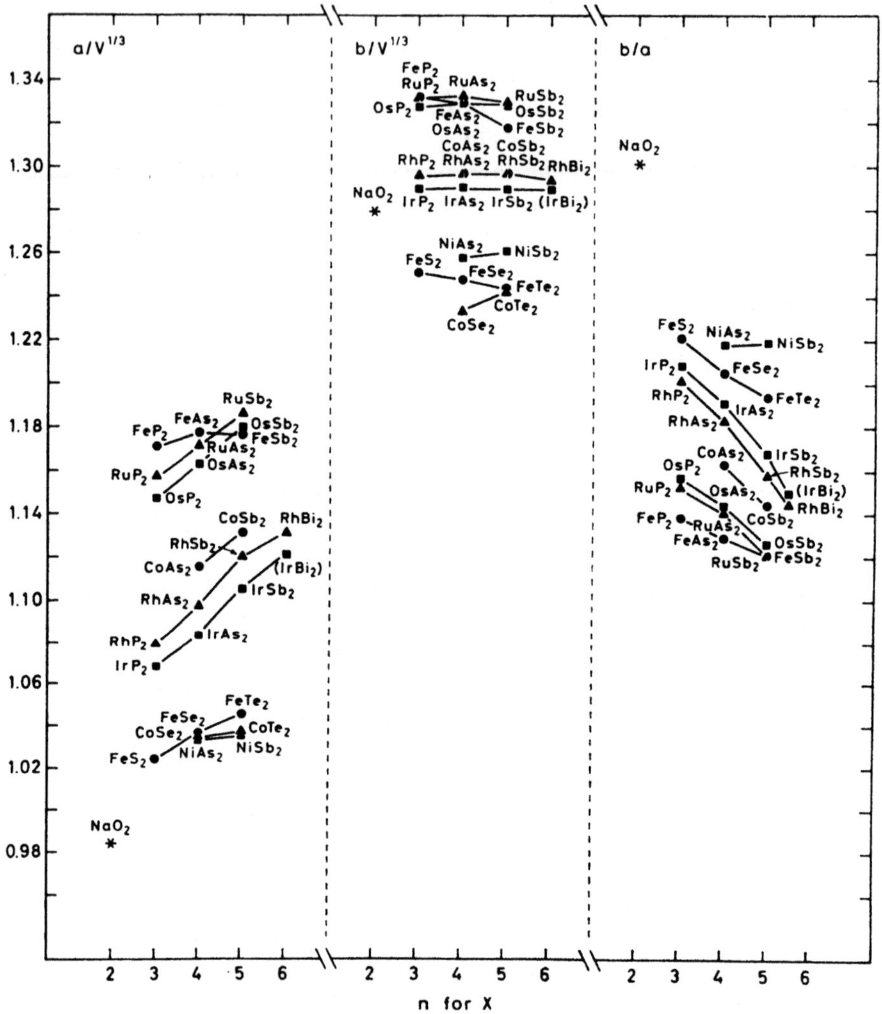
a/V^1/3
b/V^1/3
b/a
n for X
NaO2
FeP2
RuP2
OsP2
FeAs2
RuAs2
OsAs2
FeSb2
RuSb2
OsSb2
RhP2
RhAs2
RhSb2
RhBi2
IrP2
IrAs2
IrSb2
(IrBi2)
CoAs2
CoSb2
FeS2
FeSe2
FeTe2
CoSe2
CoTe2
NiAs2
NiSb2

Fig. 6a

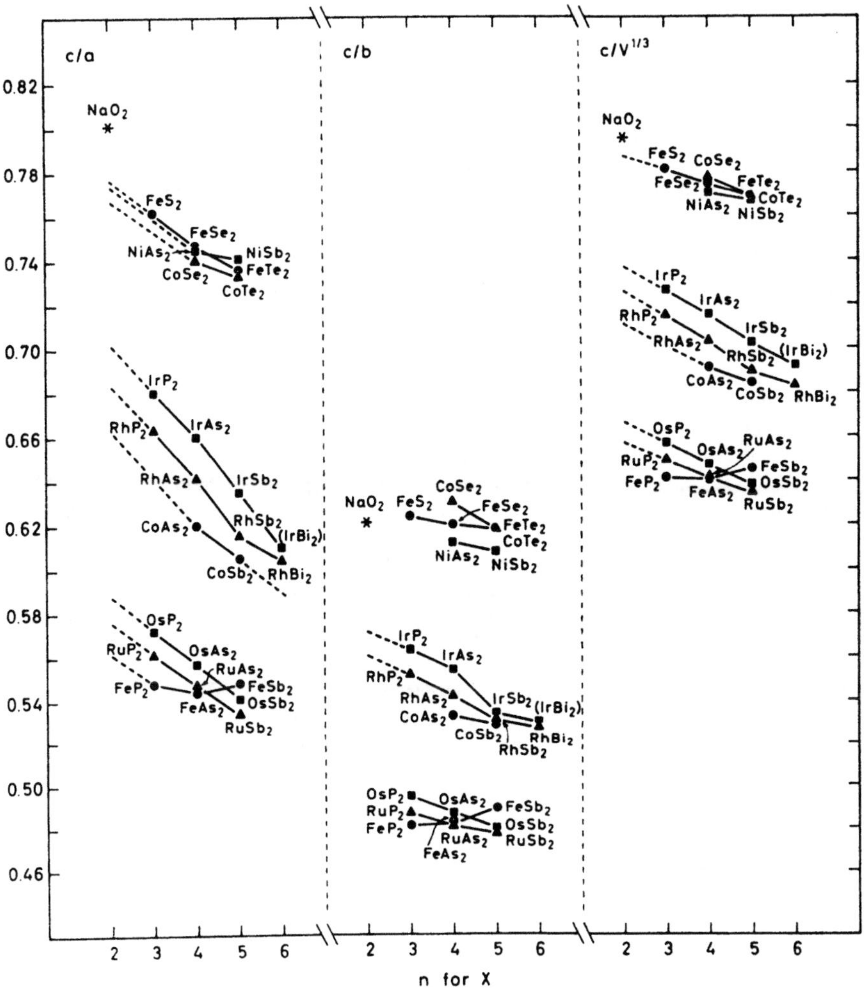

Fig. 6b

Fig. 6. Axes normalized to volume and axial ratios for compounds of the structure types FeS_2—*m* and $CoSb_2$

V. References

1. *Brostigen, G., Kjekshus, A.:* Acta Chem. Scand. *24*, 2983 (1970).
2. — — Acta Chem. Scand. *24*, 2993 (1970).
3. *Goodenough, J. B.:* J. Solid State Chem. *5*, 144 (1972).
4. *Kjekshus, A., Rakke, T.:* To be published.
5. *Brostigen, G., Kjekshus, A.:* Acta Chem. Scand. *23*, 2186 (1969).
6. — — *Rømming, Chr.:* Acta Chem. Scand. *27*, 2791 (1973).
7. *Kjekshus, A., Rakke, T., Andresen, A. F.:* Acta Chem. Scand. In press.
8. *Carter, G. F., Templeton, D. H.:* J. Am. Chem. Soc. *75*, 5247 (1953).
9. *Pearson, W. B.:* The crystal chemistry and physics of metals and alloys. New York–London–Sydney–Toronto: Wiley-Interscience 1972.
10. *Sahl, K.:* Acta Cryst. *19*, 1027 (1965).
11. *Kjekshus, A., Rakke, T.:* Acta Chem. Scand. In press.
12. *Entner, P., Parthé, E.:* Acta Cryst. *B 29*, 1557 (1973).
13. *Vannerberg, N. G.:* Acta Chem. Scand. *15*, 769 (1961).
14. *Verweel, H. J., Bijvoet, J. M.:* Z. Krist. *100*, 201 (1938).
15. *Vannerberg, N. G.:* Arkiv Kemi *14*, 119 (1959).
16. — Arkiv Kemi *10*, 455 (1956).
17. *Hoffman, C. W. W., Ropp, R. C., Mooney, R. W.:* J. Am. Chem. Soc. *81*, 3830 (1959).
18. *Vannerberg, N. G.:* Arkiv Kemi *14*, 99 (1959).
19. *Yanagisawa, S., Tashiro, M., Anzai, S.:* J. Inorg. Nucl. Chem. *31*, 943 (1969).
20. *Donohue, P. C., Siemons, W. J., Gillson, J. L.:* J. Phys., Chem. Soids *29*, 807 (1968).
21. *Bredig, M. A.:* J. Phys. Chem. *46*, 801 (1942).
22. *Atoji, M., Medrud, R. C.:* J. Phys. Chem. *31*, 332 (1959).
23. — J. Chem. Phys. *35*, 1950 (1961).
24. *Vol'nov, I. I.:* Peroxides, superoxides, and ozonides of alkali and alkaline earth metals. New York: Plenum Press 1966.
25. *Carter, G. F., Margrave, J. L., Templeton, D. H.:* Acta Cryst. *5*, 851 (1952).
26. *Vannerberg, N. G.:* Acta Chem. Scand. *16*, 1212 (1962).
27. *Bijvoet, J. M., Lely, J. A.:* Rec. Trav. Chim. *59*, 908 (1940).

Received March 22, 1974

The Electronic Spectra of the Hexafluoro Complexes of the Second and Third Transition Series

G. C. Allen* and **K. D. Warren**

Department of Chemistry, University College, Cardiff, Wales, U.K.

Table of Contents

1. Introduction

(i) Scope of Coverage

Recently (*1*) we reviewed the existing state of knowledge concerning the electronic spectra of the hexafluoro anions of the first transition series. In that account our approach was phenomenological, rather than predictive, and was largely expressed within the familiar framework of

* Present address: Central Electricity Generating Board, Berkeley Nuclear Laboratories, Berkeley, Gloucestershire, England, U.K.

the ligand field model. For this present work we have maintained our original standpoint since, although progress has been made in the molecular orbital treatment of transition metal complexes, theoretical calculations of excitation energies cannot yet be claimed to have a predictive value. Thus, for example, the treatment latterly described by *Boudreaux* (*2, 3*) for some hexafluoro anions leads to reasonable agreement with experimental data for the *d—d* transitions, but cannot be reconciled with the values reported (*1, 4*) for charge-transfer bands, unless these are arbitrarily assumed to be considerably in error. Similarly we are doubtful of the implications of the complete *ab initio* calculations which have recently been reported for some transition metal complexes [*e.g.* $CuCl_4^{2-}$ (*5*)], since these employ repulsion integrals calculated without allowance for electron correlation which are in consequence considerably greater than the experimental values, even though quite reasonable excitation energies are obtained.

Attention has also lately been drawn (*6*) to the question of the ordering of the occupied levels in the conventional molecular orbital picture [see Fig. 2 of (*1*)], and it has been suggested that the highest of these may be dominantly ligand type orbitals, rather than mostly metal type levels as was hitherto assumed. Strictly speaking this is not in any case a critical question since the *d—d* bands would nevertheless be expected (*7*) to constitute the lowest energy spectroscopic excitations, but the results of photoelectron spectra measurements (*8*) indicate that for a number of chloro complexes mainly metal orbitals do after all form the highest occupied levels. For the hexafluoro species therefore, with the lower lying ligand levels, one may reasonably assume a similar result, *i.e.* that the ordering of levels follows the traditional scheme.

Nonetheless, the electronic spectra of complexes of the second and third transition series do show a number of features which distinguish them from those of the $3d$ elements, and these we now summarise. For a given d^n configuration, representing the same oxidation state, the ligand field splitting parameter, Dq, increases quite appreciably in the sense $3d < 4d < 5d$, so that the spin-allowed *d—d* transitions occur at higher energies in the second and third series. Moreover the free-ion values of the interelectronic repulsion parameter, B, decrease in the sense $3d > 4d > 5d$ whilst the extent of covalent involvement of the ligands also diminishes, *i.e.* $\beta_{3d} > \beta_{4d} > \beta_{5d}$. On balance though smaller effective B values are found in the second and third transition series, so that those spin-forbidden excitations which arise from intra-subshell transitions tend to occur at lower energies than in the $3d$ series. In addition the increase in the nephelauxetic ratio, β, is paralleled by a corresponding decrease in the optical electronegativity, $\varkappa_{opt}$, which results in the charge-transfer transitions being located at higher energies.

The passage from the $3d$ to the $4d$ and $5d$ series is also characterised by an increase in the stability of the higher oxidation states. Thus in the $3d$ series M(IV) complexes tend to be unstable and strongly oxidising and are found only for V, Cr, Mn, Co, and Ni, and of the MF_6^{3-} compounds (Ti to Cu) those of Co, Ni, and Cu exhibit similar features. In the $4d$ series though a number of quite stable M(IV) species are known as well as examples of M(III) complexes, and MF_6^- [M(V)] anions are known for Nb, Mo, Tc, and Ru. A number of rather unstable MF_6 compounds have also been prepared, but as information concerning their electronic spectra is entirely lacking we do not treat them further here.

In the $5d$ series stable MF_6^{2-} [M(IV)] complexes are now the norm, and less stable oxidising MF_6^- anions are known for the elements Ta to Au. The neutral MF_6 molecules are also well established for the elements W to Pt, but in this case some spectroscopic information is available.

We have as yet though made no mention of the most significant feature distinguishing $4d$ and $5d$ complexes from those of the first transition series, namely the considerably greater magnitude of the spin-orbit coupling constant, ζ, within the latter. The consequences of this for the interpretation of both the *d—d* and charge-transfer spectra is discussed therefore in Section 1.(ii), and thereafter our treatment follows essentially the same pattern as in our earlier survey (*1*). In Sections 2 and 3 respectively we deal with the electronic spectra of hexafluoro compounds of the $4d$ and $5d$ series, and in Section 4 nephelauxetic and related effects are considered. Finally in Section 5 we discuss the Laporte-allowed (and other) charge-transfer bands and the related concept of optical electronegativity. We have not in this review felt it necessary to treat Jahn-Teller effects in great detail since, by virtue of the larger Dq values which here obtain, all the complexes discussed have low-spin ground states in which any orbital degeneracies arising are due to uneven occupation of the less strongly affected t_{2g} level. Moreover, such Jahn-Teller effects as may arise from the degeneracies are in most cases effectively quenched by the stronger spin-orbit coupling, leading to Jahn-Teller impotent ground states.

(ii) Basic Theory

As indicated in Section 1.(i) our approach to the interpretation of the electronic spectra of the hexafluorometallates of the $4d$ and $5d$ series closely follows that employed previously (*1*), and need not be recapitulated in detail. The ligand field model is adopted for the interpretation of the *d—d* transitions, and the qualitative molecular orbital approach for the treatment of both these and, more particularly, the charge-transfer excitations. Because of the larger Dq and smaller B values encountered in the second and third transition series it is clear that the strong field version of the ligand field model will constitute the only realistic approach for the interpretation of the d—d spectra. However, although spin-orbit interactions are rather small in the $3d$ series, and usually need not be included explicitly in any calculations, ζ becomes considerably greater for the $4d$ and $5d$ elements. Thus for the $3d$ series the spin-orbit coupling constant ranges between about 100 and 800 cm^{-1}, but for the $4d$ and $5d$ blocks the corresponding values are respectively 300—1800 cm^{-1}m and 1300—5000 cm^{-1}. Specific inclusion of spin-orbit effects is therefore obviously essential for $5d$ complexes, and preferable for those of the $4d$ elements. Accordingly, in assigning and fitting the spectra which we ourselves have reported (*9—12*) we have always incorporated the spin-orbit terms, utilising for this purpose the strong field matrices given by *Liehr* and *Ballhausen* (*13*) — d^2, d^8 —, by *Eisenstein* (*14, 15*) — d^3, d^7 —, and by *Schroeder* (*16, 17*) — d^4, d^6 and d^5.

Such inclusion of spin-orbit interactions is however of value quite apart from the calculation of band positions, and splittings due to spin-orbit coupling. Thus the eigenvectors obtained by diagonalisation of the complete strong field energy matrices (with appropriate parameters) can be used to make estimates of the intensities of the spin-forbidden bands relative to those of the spin-allowed transitions, and comparison of these with the experimental results can prove of considerable assistance in making assignments. In general such calculations turn out to be in accord with the observed tendency for the relative intensities of the spin-forbidden excitations to increase on passing from the $3d$ to the $4d$ and to the $5d$ series.

Although, as shown above, it is always desirable for spin-orbit terms to be included, it is not always possible for this to be done very precisely. Thus, in the $4d$ series, ζ, although appreciable, is often not large enough for any first order band splittings to be observed, and an approximate value, *e.g.* 1000 cm^{-1} as in (*12*), may have to be used. However, in the $5d$ series, the much larger value of ζ frequently results in observable spin-orbit band splittings, and for d^4 and d^5 systems, with $^3T_{1g}$ and $^2T_{2g}$ ground states respectively, the direct observation of the spin-orbit splitting of the ground state manifold enables quite good estimates of the appropriate effective ζ values to be made for the complexes concerned. Under such circumstances, if values of ζ for the free metal ion can be obtained, the extent of reduction of the spin-orbit coupling constant due to covalency can be assessed. One may thus define a quantity, β^*, the relativistic ratio, equal to $\zeta_{complex}/\zeta_{free-ion}$, and this, in conjunction with the nephelauxetic ratio β $(=B_{complex}/B_{free-ion})$, may be used to make estimates of the extents of central field and symmetry restricted covalency. Details of this, and discussion of the factors determining the magnitude of β^*, are given in Section 4. (i).

Spin-orbit (or relativistic) contributions are also important for the understanding of charge-transfer spectra, and must be taken into account in the calculation of optical electronegativities, $\varkappa_{opt}$, from the positions, σ_{obs}, of the first Laporte-allowed bands. In addition to the corrections to σ_{obs} listed before (*1*), allowance should in principle also be made for any changes in spin-orbit contributions between the d^q and d^{q+1} ground states. In the $3d$ series such corrections are negligible, and will not amount to more than about 0.05 $\varkappa_{opt}$ unit for the $4d$ complexes: however, for $5d$ species substantial corrections to σ_{obs}, due to relativistic effects may ensue (*10*, *11*), and the inclusion of these terms enables a better rationalisation to be made for the optical electronegativity data for a wide range of $5d$ systems (*18*). A full discussion of these effects is given in Section 5, in which relativistically corrected optical electronegativities, $\overset{*}{\varkappa}_{opt}$, are defined, and the appropriate spin-orbit contributions listed for the various t_{2g}^q ground states.

2. Electronic Spectra of the Hexafluoro Species of the $4d$ Series

(i) MF_6^{3-} Systems

Unlike the trivalent cations of the first transition series, those of the $4d$ elements give rise only to a limited number of MF_6^{3-} anions. Thus, instead of all the configurations from d^1 to d^8 being accessible, only those derived from Mo(III) (d^3), Ru(III) (d^5), Rh(III) (d^6), and Ag(III) (d^8), are known as hexafluoro complexes, thereby underlining the increasing stability of the higher oxidation states. Furthermore, despite the generally weak nephelauxetic character of the fluoride ion, the larger Dq values and smaller B values in the $4d$ series ensure that low-spin ground states are found for RuF_6^{3-} and RhF_6^{3-}, in contrast to the high-spin behaviour shown by the corresponding $3d$ complexes, FeF_6^{3-} and CoF_6^{3-}. As in the $3d$ series though the B values derived from fitting of the d–d spectra indicate a decreasing trend in β, the nephelauxetic ratio, and thereby an increasing tendency to covalency and diminishing stability, towards the end of the transition series.

a) *Hexafluoromolybdate(III)*, MoF_6^{3-}

The hexafluoromolybdate(III) anion was first obtained as the potassium salt, K_3MoF_6, by *Peacock* (*19*), via the fusion of K_3MoCl_6 with KHF_2. A brown cubic latticed product was produced with a magnetic moment of 3.2 B.M. at 298 °K, but no spectroscopic data were reported. More recently though K_3MoF_6 was prepared by *Toth, Brunton,* and *Smith* (*20*) by the fusion of KF with MoF_3, and an X-ray diffraction study was described which established a cubic cryolite family lattice for the lemon yellow product.

These latter authors though also reported a diffuse reflectance spectrum for their compound, which showed absorption peaks at 23.5, 29.7, and 38.2 kK. (See Fig. 1); the first two bands were assigned as the $^4A_{2g} \rightarrow {}^4T_{2g}$ and $^4A_{2g} \rightarrow {}^4T_{1g}$ (F) *d—d* transitions respectively, as shown in Table 1, from which the parameters $Dq = 2350$ cm^{-1} and $B = 570$ cm^{-1} were deduced. The limited range of the instrument employed (16—50 kK.) precluded any attempt to detect the expected lower energy spin-forbidden transitions, $^4A_{2g} \rightarrow {}^2E_g$, $^4A_{2g} \rightarrow {}^2T_{1g}$, and $^4A_{2g} \rightarrow {}^2T_{2g}$, but the band at 38.2 kK. is 11 kK. lower than the Dq and B parameters predict for the higher energy $^4A_{2g} \rightarrow {}^4T_{1g}$ (P) transition, and seems to be too strong to correspond to the spin-forbidden excitations anticipated in the region of 40 kK. *Toth et al.* (*20*) ascribe the band to a charge-transfer transition, but the known optical electronegativity for Mo(III) (see Section 5) indicates that the lowest allowed $\pi \rightarrow t_{2g}$ band should not occur below about 60 kK. The spectrum is not however particularly well resolved, and the intensities are expressed only in terms of relative absorbance, rather than the *Kubleka-Munk* F_R values, so that too great a reliance should not be placed on the intensity of the 38.2 kK. band. Consequently the assignment of this band to a group of spin-forbidden transitions cannot be altogether excluded, and the rising absorption observed at about 43.5 kK. accords quite well with the anticipated position of the $^4A_{2g} \rightarrow {}^4T_{1g}$ (P) peak. A more detailed study of the spectrum of this compound, over a wider energy range, is though clearly desirable.

Table 1. Spectroscopic data for potassium hexafluoromolybdate (III)

Band Position (kK.)	Assignment
23.5	$^4A_{2g} \rightarrow {}^4T_{2g}$
29.7	$^4A_{2g} \rightarrow {}^4T_{1g}$ (F)
38.2	charge-transfer[a]) or spin-forbidden bands (?)[b])

$Dq = 2350$ cm^{-1}, $B = 570$ cm^{-1}

[a]) Ref. (*20*). [b]) Present work.

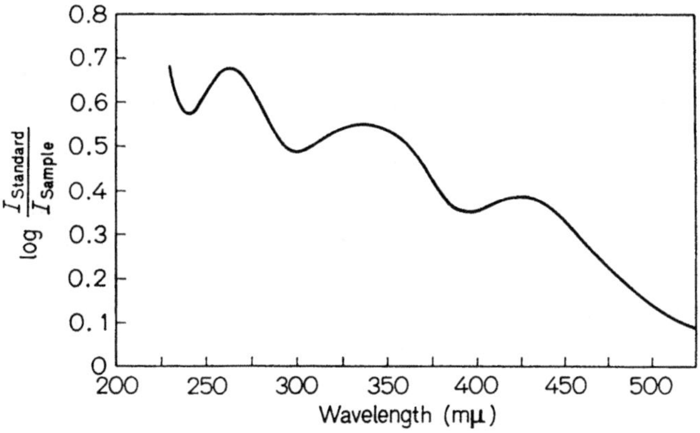

Fig. 1. Electronic spectrum of K_3MoF_6

b) Hexafluororuthenate(III), RuF_6^{3-}

Potassium hexafluororuthenate(III) is obtained as a dark solid by fusion of $RuCl_3$ with KHF_2 in a stream of nitrogen, according to the method of *Peacock* (*19*, *21*). A magnetic moment of 1.25 B.M. at 298 °K has been reported (*19*) for the anion, thus clearly implying a low-spin $^2T_{2g}$ ($t_{2g}{}^5$) ground state. The electronic spectrum of K_3RuF_6 has been studied by *Allen*, *El-Sharkawy*, and *Warren* (*12*) by diffuse reflectance over the range 4–50 kK., and the band positions and their assignments are given in Table 2. The observed spectrum is shown in Fig. 2.

The weak bands near 10 and 15.4 kK. were identified as the spin-forbidden $^2T_{2g} \rightarrow {}^4T_{1g}$ and $^2T_{2g} \rightarrow {}^4T_{2g}$ transitions respectively, and the stronger bands at 20.0, 26.5, and 34.0 kK. assigned as arising from groups of spin-allowed $t_{2g}^5 \rightarrow t_{2g}^4\, e_g$ transitions. The absorption finally rises sharply just below 50 kK. and the presence of a *Laporte* allowed , $\pi \rightarrow t_{2g}$ transition in the region of 55 kK. was inferred. With a C/B ratio of 4.75 (see Section 4. (i)), the *d—d* bands were best fitted by the parameters $Dq = 2200$ cm^{-1} and $B = 550$ cm^{-1}: for the reasons outlined in Section 1. (ii) it was not possible to obtain an accurate figure for the effective value of ζ, but, on the basis of the free-ion value, the approximation $\zeta = 1000$ cm^{-1} was adopted for the calculations.

In the region of 4 kK. some slight absorption was noticed, but no clear indication of a band. Spin-orbit coupling splits the $^2T_{2g}$ (t_{2g}^5) d^5 ground state into an upper Γ_8 and a lower Γ_7 component, separated by (3/2) ζ in the first order, so that a low energy $^2T_{2g}$ (Γ_7) $\rightarrow {}^2T_{2g}$ (Γ_8) transition should occur at about 1.5 kK. Diagonalisation of the full strong field energy matrices gives a value of 1.9 kK., but it seems that the effective ζ value ($\zeta_{\text{free-ion}}$, Ru(III) = 1180 cm^{-1}) is not great enough for the tail of any such band to be seen near 4 kK., the lower limit of measurement.

For the low lying $^4T_{1g}$ and $^4T_{2g}$ levels the overall spin-orbit splitting is predicted to be the same in the first order, but here diagonalisation of the energy matrices suggests a spread of some 1.4 kK. for the former, and only 0.3 kK. for the latter (see Table 2). The experimental spectrum however bears out these calculations since the $^2T_{2g} \rightarrow {}^4T_{1g}$ band is broad and ill defined, whilst the $^2T_{2g} \rightarrow {}^4T_{2g}$ absorption is a much narrower peak.

For low-spin d^5 ground states a large number of overlapping spin-allowed $t_{2g}^5 \rightarrow t_{2g}^4\, e_g$ bands are found at higher energies: the sharp peak at 20.0 kK. is reasonably assigned to three juxtaposed transitions, $^2T_{2g} \rightarrow {}^2A_{2g}$, $^2T_{1g}$, $^2T_{2g}$, but beyond this the band attributions become somewhat precarious. The assignments listed in Table 2 must therefore be regarded as slightly tentative.

Table 2. Spectroscopic data for potassium hexafluororuthenate (III)

Band Position (kK.)		Assignment
Obsd.	Calcd.[a]	
	10.3 (Γ_8)	
	10.5 (Γ_6)	
10.0	11.5 (Γ_8)	$^2T_{2g} \rightarrow {}^4T_{1g}$
	11.7 (Γ_7)	
	15.3 (Γ_7)	
	15.4 (Γ_8)	
15.4	15.6 (Γ_8)	$^2T_{2g} \rightarrow {}^4T_{2g}$
	15.6 (Γ_6)	
20.0	20.4, 20.4, 22.1	$^2T_{2g} \rightarrow {}^2A_{2g}, {}^2T_{1g}, {}^2T_{2g}$
26.5 (sh)	23.4, 27.0, 28.8	$^2T_{2g} \rightarrow {}^2E_g, {}^2T_{1g}, {}^2T_{2g}$
34.0 (br)	30.4, 35.2	$^2T_{2g} \rightarrow {}^2A_{1g}, {}^2E_g$
50.0	—	$\pi \rightarrow t_{2g}$

[a]) Dq = 2200 cm^{-1}, B = 550 cm^{-1}, ζ = 1000 cm^{-1}, C/B = 4.75.

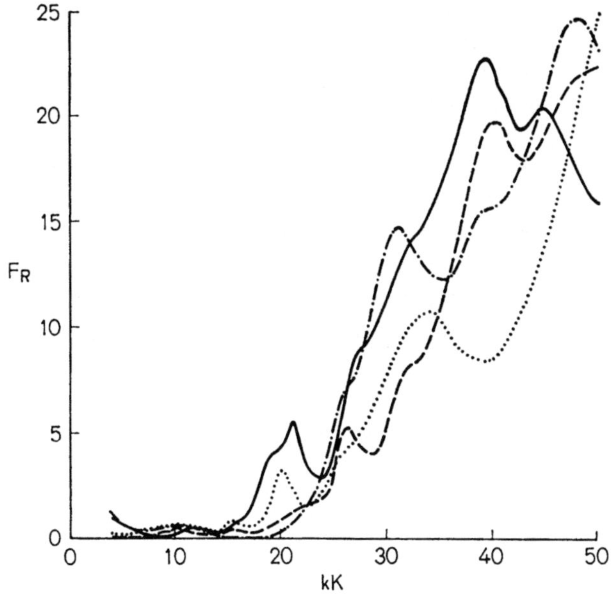

Fig. 2. Electronic spectra of K_3RuF_6, $KRuF_6$, and Cs_2RhF_6

c) *Hexafluororhodate*(III), RhF_6^{3-}

Potassium hexafluororhodate(III), K_3RhF_6, was obtained by *Peacock* (*22*) as a buff solid by fusion of $K_3Rh(NO_2)_6$ with KHF_2. It was found to be diamagnetic (*19*), thus implying the presence of a low-spin ${}^1A_{1g}$ (t_{2g}^6) ground state. The electronic spectrum was studied by diffuse reflectance by *Schmidtke* (*23*) over the range 15—45 kK., as shown in Fig. 3, and the bands observed are listed in Table 3.

At lower energies two peaks were observed at 15.5 and 16.5 kK. and these collectively were attributed to the transitions ${}^1A_{1g} \rightarrow {}^3T_{1g}$, ${}^3T_{2g}$. At higher energy two well marked peaks of moderate intensity were found at 21.3 and 27.8 kK., which clearly correspond to the ${}^1A_{1g} \rightarrow {}^1T_{1g}$ and ${}^1A_{1g}$, $t_{2g}^6 \rightarrow t_{2g}^5\, e_g$, bands, and yield the parameters $Dq = 2230$ cm^{-1} and $B = 460$ cm^{-1} above this the absorption rises slowly, but without any indication of a Laporte-allowed charge-transfer band below 45 kK., and in consequence the small inflection at 37.0 kK. was tentatively attributed to a two-electron excitation, ${}^1A_{1g} \rightarrow {}^1T_{2g}$ ($t_{2g}^4\, e_g^2$). Indeed, optical electronegativity considerations (see Section 5) suggest strongly that no charge-transfer transitions should be observable here since for $RhCl_6^{3-}$ the $\pi \rightarrow e_g$ excitation already lies as high as 39.2 kK. (*24*).

The assignment of the spin-forbidden transitions does however merit some comment. To the first order the $^1A_{1g} \to {}^3T_{1g}$ and $^1A_{1g} \to {}^3T_{2g}$ transitions should be separated by 8 B, that is some 3.5 kK. Clearly therefore the spacing of 1kK. between the two weak peaks is much too small to permit their assignment in this way. Moreover, taking C/B as 4.8, the same value as found by *Tanabe* and *Sugano* (*25*) for the corresponding 3*d* oxidation state, Co(III), the $^1A_{1g} \to {}^3T_{1g}$ peak is predicted to lie in the region of 16—17 kK., and the $^1A_{1g} \to {}^3T_{2g}$ band near 20 kK., *i.e.* where it would be obscured by the broad spin-allowed $^1A_{1g} \to {}^1T_{1g}$ absorption. A more likely assignment of the 15.5 and 16.5 kK. absorptions seems therefore to be to ascribe them to spin-orbit split components

Table 3. Spectroscopic data for potassium hexafluororhodate (III)

Band Position (kK.)	Assignment
15.5	
16.5	$^1A_{1g} \to {}^3T_{1g}, {}^3T_{2g}$[a]) or $^1A_{1g} \to {}^3T_{1g}$[b])
21.3	$^1A_{1g} \to {}^1T_{1g}$
27.8	$^1A_{1g} \to {}^1T_{2g}$

$Dq = 2230$ cm^{-1}, $B = 460$ cm^{-1}, $C/B = 4.8$.
[a]) Ref. (*23*). [b]) Present work.

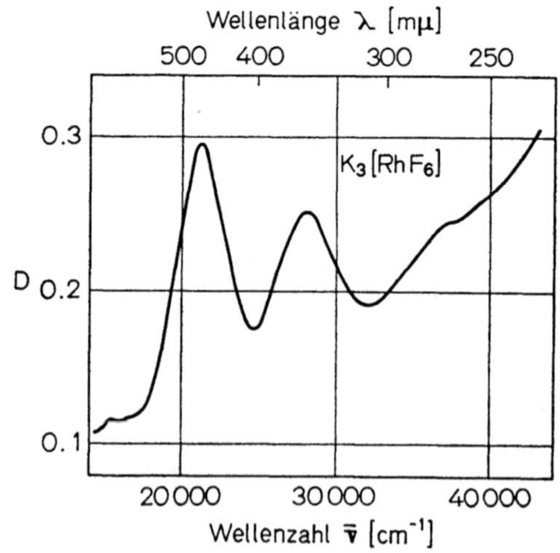

Fig. 3. Electronic spectrum of K_3RhF_6

of the $^3T_{1g}$ level: the overall first order splitting amounts to (3/4) ζ, which, taking ζ Rh(III) 1300 cm^{-1}, is consistent with the observed data. Extension of the measurements to lower energies would here be valuable so as to establish the absence (or otherwise) of any lower energy spin-forbidden bands.

d) *Hexafluoroargentate*(III), AgF_6^{3-}

The hexafluoroargentate(III) anion was first obtained as the Cs_2K salt by *Hoppe* and *Homann* (*26*), who fluorinated a 2:1:1 mixture of CsCl, KCl, and $AgNO_3$ at 300 °C. A moment of 2.6 B.M. was reported for the product, and the electronic spectrum was studied by *Allen* and *Warren* (*9*), using the diffuse reflectance technique.

A weak peak was found at 6.3 kK. together with a small shoulder at 12.9 kK., followed by stronger bands at 18.4 and 23.4 kK.. The first two of these were assigned as the spin-forbidden excitations, $^3A_{2g} \rightarrow {^1E_g}$ and $^3A_{2g} \rightarrow {^1A_{1g}}$ respectively, and the latter two to the spin-allowed *d—d* transitions $^3A_{2g} \rightarrow {^3T_{2g}}$ and $^3A_{2g} \rightarrow {^3T_{1g}}$. From the spin-allowed bands one obtains $Dq = 1840$ cm^{-1} and $B_{35} = 472$ cm^{-1}, whilst the spin-forbidden levels yield $B_{33} = 375$ cm^{-1}. This observation, $B_{33} < B_{35}$, is in agreement with the expectation of greater covalent involvement of the σ-bonding e_g level, as compared with the π-bonding t_{2g} orbital. Above the 23.4 kK. band a region of more intense absorption supervenes, with peaks at 27.5 and 37.5 kK. which were attributed to $\pi \rightarrow e_g$ charge-transfer transitions (see Section 5).

For AgF_6^{3-} the intensities of the spin-forbidden bands, relative to those of the allowed transitions, are actually smaller than for the analogous 3*d* complex, CuF_6^{3-}. This however is because the larger *Dq* of the 4*d* series, and the smaller *B* value, results in a greater separation between

Table 4. Spectroscopic data for dicaesium potassium hexafluoroargentate (III)

Band Position (kK.)	Assignment
6.3	$^3A_{2g} \rightarrow {^1E_g}$
12.9	$^3A_{2g} \rightarrow {^1A_{1g}}$
18.4	$^3A_{2g} \rightarrow {^3T_{2g}}$
23.4	$^3A_{2g} \rightarrow {^3T_{1g}}$
27.5	$\pi \rightarrow e_g$
37.5	$\pi \rightarrow e_g$

$Dq = 1840$ cm^{-1}, $B_{35} = 470$ cm^{-1}, $B_{33} = 375$ cm^{-1}, $\zeta = 1700$ cm^{-1}, $C/B = 5.0$.

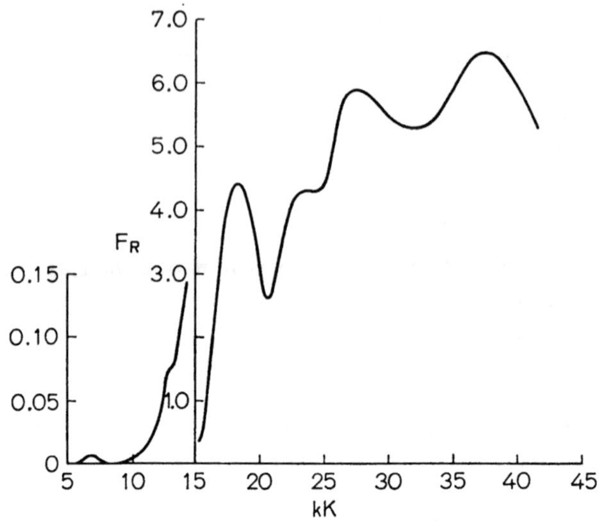

Fig. 4. Electronic spectrum of Cs_2KAgF_6

the triplet ($t_{2g}^5\ e_g^3$) and singlet ($t_{2g}^6\ e_g^2$) levels: thus, despite the larger ζ for the $4d$ metal the extent of spin-orbit mixing in this case is appreciably smaller. (See Ref. 9 for detailed calculations, taking $\zeta \sim 1700\ cm^{-1}$.)

(ii) MF_6^{2-} Systems

In the $4d$ series the M(IV) oxidation state represents one of the more common and stable valencies. Hexafluoroanions, MF_6^{2-}, are known for all the elements from Mo to Pd, but the general trend of decreasing stability towards the end of the series is manifested here too. As for the M(III) species, low-spin ground states are found throughout, including the d^4 complex, RuF_6^{2-}, for which no analogue is known in the MF_6^{3-} series. For the MoF_6^{2-} anion [obtained by the action of excess NaI on MoF_6 in liquid SO_2 (*27*)] no electronic spectrum has been reported, but for the remaining species satisfactory experimental data are available and are surveyed below.

a) Hexafluorotechnetate(IV), TcF_6^{2-}

Potassium hexafluorotechnetate(IV), K_2TcF_6, is obtained from potassium pertechnetate via the hexabromo derivative with subsequent fusion with KHF_2 (*28*), and the electronic spectrum (see Table 5) has been measured in aqueous solution by *Jørgensen* and *Schwochau* (29). Two groups of weak bands were found near 11 and 18 kK., the peak at 11.0 and the shoulder at 11.5 kK. being assigned as ${}^4A_{2g} \rightarrow {}^2E_g$, ${}^2T_{1g}$, and that at 17.6 kK. (with a shoulder at 18.1 kK.) as ${}^4A_{2g} \rightarrow {}^2T_{2g}$. Both of these groups of peaks show fairly narrow band widths as would be expected for intra-subshell transitions within the t_{2g} manifold (see Fig. 5). At higher energies two broader bands of moderate intensity were found at 28.4 and 34.4 kK., and assigned as the ${}^4A_{2g} \rightarrow {}^4T_{2g}$ and ${}^4A_{2g} \rightarrow {}^4T_{1g}$ (F), $t_{2g}^3 \rightarrow t_{2g}^2\ e_g$, transitions respectively. A somewhat stronger peak was found at 46.3 kK., but this appears to lie at too low an energy to correspond to the ${}^4A_{2g} \rightarrow {}^4T_{1g}$ (P) transition: it also seems to be too low to represent a Laporte-allowed charge-transfer transition, since the position of the $\pi_u \rightarrow t_{2g}$ band in $TcCl_6^{2-}$ ($\sim$30 kK.) would predict a value of 55—60 kK. for this excitation. (See Section 5). From the spin-allowed bands the parameters $Dq = 2840\ cm^{-1}$ and $B_{35} = 530\ cm^{-1}$ were derived, whilst the spin-forbidden transitions yielded $B_{55} = 555\ cm^{-1}$. The relative magnitudes of the B_{55} and B_{35} values is as would be expected on account of the generally smaller covalent involvement of the π-bonding t_{2g} level.

It is though noteworthy that here again the spin-forbidden bands are actually slightly weaker in intensity, relative to the spin-allowed transitions, than in the $3d$ analogue, MnF_6^{2-}, [see (1)]. This is because although the free ion ζ value for Tc(IV) is about twice that for Mn(IV) (ca. 850 cm^{-1} as against ca. 400 cm^{-1}), the ${}^2E_g$ and ${}^2T_{1g}$ levels lie significantly further below the quartet levels and are therefore less extensively mixed by the spin-orbit interaction. The magnitude of ζ Tc(IV) is not however sufficient to cause any observable splittings of the quartet levels whose overall separations should both amount only to $2/3\ \zeta$, *i.e.* $\sim$ 600 cm^{-1}, in the first order.

Table 5. Spectroscopic data for potassium hexafluorotechnetate (IV)

Band Position (kK.)	Assignment
11.0	${}^4A_{2g} \rightarrow {}^2E_g$
11.5	${}^4A_{2g} \rightarrow {}^2T_{1g}$
17.6 18.1 (sh)	${}^4A_{2g} \rightarrow {}^2T_{2g}$
28.4	${}^4A_{2g} \rightarrow {}^4T_{2g}$
34.4	${}^4A_{2g} \rightarrow {}^4T_{1g}$ (F)
46.3	?; $\pi_g \rightarrow t_{2g}$[a])

$Dq = 2840\ cm^{-1}$, $B_{35} = 530\ cm^{-1}$, $B_{55} = 555\ cm^{-1}$.
[a]) Present work.

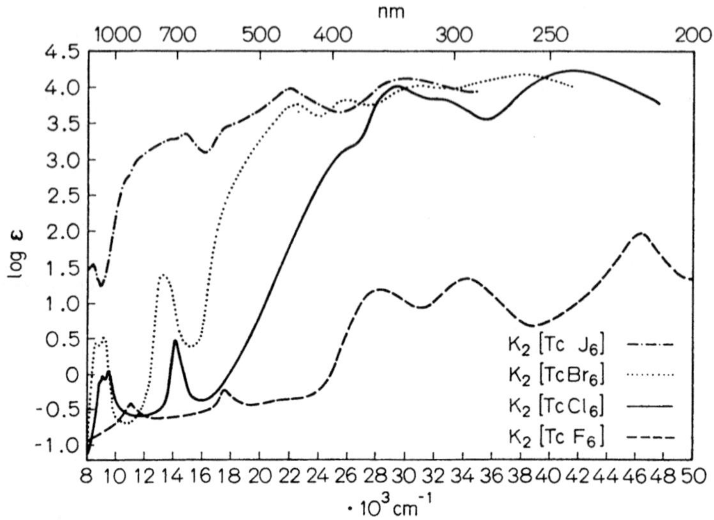

Fig. 5. Electronic spectrum of K_2TcF_6

b) Hexafluororuthenate(IV), RuF_6^{2-}

Potassium hexafluororuthenate(IV), K_2RuF_6, is readily obtained by hydrolysis of $KRuF_6$ (q.v.) (*30*). It has been found (*30, 31*) to have a magnetic moment of 2.86 B.M. at 298 °K and therefore clearly possesses a low-spin $^3T_{1g}$, (t_{2g}^4), ground state. Two studies of the diffuse reflectance spectrum have been reported, the first by *Brown, Russell* and *Sharp* (*32*), and a more recent investigation by *Allen, El-Sharkawy*, and *Warren* (*12*). *Brown et al.* studied a number of $4d$ and $5d$ hexafluoro species over the limited range 10—40 kK., but the results were not presented diagrammatically, and specific allowance for spin-orbit effects was omitted from their interpretations. For K_2RuF_6 they reported only two bands, at 27 and 31 kK., assigned as *d—d* transitions, but the results of *Allen et al.* between 4 and 50 kK. reveals a much richer spectrum. (See Table 6 and Fig. 2).

A broad rather weak band was found just below 10 kK. and assigned to the closely grouped spin-forbidden transitions $^3T_{1g} \rightarrow {}^1T_{2g}, {}^1E_g, {}^5E_g$, whilst at higher energies stronger bands at 26.8 and 31.0 kK. (c.f. *Brown et al.*) were attributed to groups of closely juxtaposed spin-allowed, $t_{2g}^4 \rightarrow t_{2g}^3\, e_g$, *d—d* transitions. Above this a rising absorption showed a band at 38.8 kK., also ascribed to a *d—d* transition, and an intense peak at 48.0 kK., assigned as a Laporte-allowed $\pi \rightarrow t_{2g}$ charge-transfer excitation.

Just above 4 kK. the beginning of a moderately strong band at lower energy was detected and assigned as a transition, ${}^3T_{1g}$ $(\Gamma_1) \rightarrow {}^3T_{1g}$ (Γ_3, Γ_5) between the spin-orbit split components of the ground state. For a ${}^3T_{1g}$ (t_{2g}^4) ground state when spin-orbit coupling is admitted the double group levels Γ_1, Γ_3, Γ_4, and Γ_5 arise, of which Γ_1 lies lowest and the first-order degenerate Γ_3, Γ_5 levels highest. The total first-order splitting is equal to $3/2\ \zeta$, but with $\zeta \sim 1000$ cm^{-1} a full calculation predicted a value of 1.8 kK. for this excitation. Thus, allowing for the co-excitation of one quantum of the enabling vibration the peak would be expected to lie between about 2.0 and 2.5 kK., and since similar transitions are known to lead to quite broad bands in the 5*d* series, the detection of the tail of this absorption near 4 kK. is not unexpected.

From their spectrum *Allen et al.* derived the parameters $Dq = 2500$ cm^{-1}, and $B = 500$ cm^{-1}, assuming $\zeta = 1000$ cm^{-1} and $C/B = 4.75$. A somewhat better fitting might have been possible if C and B had been allowed to vary independently but for reasons explained in Section 4.(i) this was not permitted.

For ruthenium and the following elements of the 4*d* series the free-ion ζ values become in excess of 1000 cm^{-1}, and apart from the exceptions mentioned previously, the relative intensities of the spin-forbidden bands are found to be significantly larger than for the 3*d* elements. For the intra-subshell (t_{2g}^4) transitions ${}^3T_{1g} \rightarrow {}^1T_{2g}$ and ${}^3T_{1g} \rightarrow {}^1E_g$ some 2% of triplet character is calculated for the singlet levels, so that when allowance is made for the narrow band width expected for such transitions, the observed relative intensities of the spin-forbidden and spin-allowed *d–d* transitions is very much of the correct order of magnitude to cor-

Table 6. Spectroscopic data for potassium hexafluororuthenate (IV)

Band Position (kK.)		Assignment
Obsd.	Calcd.[a]	
4.0	1.7 (Γ_3, Γ_5)	${}^3T_{1g}$ $(\Gamma_1) \rightarrow {}^3T_{1g}$ (Γ_3, Γ_5)
	8.2	${}^3T_{1g} \rightarrow {}^1T_{2g}$
9.8 (br)	8.8	${}^3T_{1g} \rightarrow {}^1E_g$
	12.0—12.1	${}^3T_{1g} \rightarrow {}^5E_g$
26.8 (sh)	24.0, 24.7, 25.3	${}^3T_{1g} \rightarrow {}^3E_g, {}^3T_{1g}, {}^3T_{2g}$
31.0	26.0, 27.0, 28.2	${}^3T_{1g} \rightarrow {}^3A_{1g}, {}^3A_{2g}, {}^3E_g$
38.8	35.3, 35.6	${}^3T_{1g} \rightarrow {}^3T_{1g}, {}^3T_{2g}$
48.0	—	$\pi \rightarrow t_{2g}$

[a]) $Dq = 2500$ cm^{-1}, $B = 500$ cm^{-1}, $\zeta = 1000$ cm^{-1}, $C/B = 4.75$.

respond to the experimental results (*12*). Since however the $^3T_{1g} \rightarrow {^5E_g}$ excitation corresponds to a $t_{2g}^4 \rightarrow t_{2g}^3e_g$ transition it probably makes only a minor contribution to the intensity of the 9.8 kK. band.

c) *Hexafluororhodate* (IV), RhF_6^{2-}

Caesium hexafluororhodate (IV), Cs_2RhF_6, is obtained by the action of BrF_3 on a 2:1 mixture of CsCl and $RhCl_3$ following the method described by *Sharpe* (*33*) and by *Peacock* (*19*). A magnetic moment of 1.7—1.9 B.M. has been reported for the Na salt (*34*) so that the RhF_6^{2-} ground state is obviously the low-spin $^2T_{2g}$ (t_{2g}^5).

Once again two spectroscopic studies have been made — due to *Brown et al.* (32) and to *Allen et al.* (12) respectively — and the same comments apply as for the two investigations of the RuF_6^{2-} anion. *Brown et al.* reported spin-forbidden bands at 12.2 and 16.1 kK., with spin-allowed transitions at 19—21 kK. and at 26.0 kK., but no other absorptions were found below 40 kK. On the other hand the spectrum of *Allen et al.*, although broadly mirroring these findings, revealed extra absorptions at 32.8, 39.6, and 44.6 kK., together with a strong indication of a band below 4 kK. (Fig. 2).

As for RuF_6^{2-} this low energy band was attributed to an intra-ground state transition between the spin-orbit split components. In O* the $^2T_{2g}$ (t_{2g}^5) state yields the components Γ_7 and Γ_8, the former lying lower, with a first order splitting of 3/2 ζ. For the $^2T_{2g}$ (Γ_7) $\rightarrow$ $^2T_{2g}$ (Γ_8) transition full calculations gave a value of 1.8 kK. so that allowing for vibronic coupling the band maximum would be expected to lie in the 2.0—2.5 kK. region and to show a detectable absorption near 4 kK. (c.f. RuF_6^{2-}). At higher energies *Allen et al.* reported weak bands at 11.6 and 16.0 kK., assigned as the spin-forbidden transitions $^2T_{2g} \rightarrow {^4T_{1g}}$ and $^2T_{2g} \rightarrow {^4T_{2g}}$ respectively. As for the iso-electronic RuF_6^{3-} (q.v.) the calculated splitting of the $^4T_{1g}$ band was noticeably larger than that of the $^4T_{2g}$ level, and the experimental spectrum is consistent with this prediction. Above these spin-forbidden bands the stronger 19—21 kK. absorption was resolved into two components, at 19.2 and 21.2 kK., which were assigned as spin-allowed *d—d* transitions, $^2T_{2g} \rightarrow {^2T_{1g}}$, $^2A_{2g}$ and $^2T_{2g} \rightarrow {^2T_{2g}}$, ^2E_g. At higher energies again a large number of further closely adjacent $t_{2g}^5 \rightarrow t_{2g}^4\, e_g$ bands are anticipated whose assignment was necessarily tentative. Shoulders at 27.6 and 32.8 kK. were thus attributed to $^2T_{2g} \rightarrow {^2T_{1g}}$, $^2T_{2g}$, $^2A_{1g}$ and $^2T_{2g} \rightarrow {^2E_g}$ excitations, and the intense broad band with peaks at 39.6 and 44.6 kK. to Laporte-allowed charge-transfer transitions. This latter assignment was noted as being consistent with the known optical electronegativity of Rh(IV).

The *d—d* bands were fitted, using the parameters $Dq = 2250\ cm^{-1}$, $B = 410\ cm^{-1}$, and $\zeta = 1000\ cm^{-1}$, with $C/B = 4.90$.

Table 7. Spectroscopic data for caesium hexafluororhodate(IV)

Band Position Γ (kK.) Obsd.	Calcd.[a])	Assignment
4.0	1.8 (Γ_8)	$^2T_{2g}(\Gamma_7) \rightarrow {}^2T_{2g}(\Gamma_8)$
	11.9 (Γ_8)	
	12.2 (Γ_7)	
11.6	12.6 (Γ_8)	$^2T_{2g} \rightarrow {}^4T_{1g}$
	12.9 (Γ_6)	
	15.5 (Γ_6)	
	15.5 (Γ_8)	
16.0	15.8 (Γ_8)	$^2T_{2g} \rightarrow {}^4T_{2g}$
	15.9 (Γ_7)	
19.2	19.4, 19.5	$^2T_{2g} \rightarrow {}^2T_{1g}, {}^2A_{2g}$
21.2	20.8, 21.8	$^2T_{2g} \rightarrow {}^2T_{2g}, {}^2E_g$
27.6 (sh)	24.5, 25.9, 27.2	$^2T_{2g} \rightarrow {}^2T_{1g}, {}^2T_{2g}, {}^2A_{1g}$
32.8 (sh)	31.6	$^2T_{2g} \rightarrow {}^2E_g$
39.6	—	$\pi \rightarrow t_{2g}$
44.6	—	$\pi \rightarrow t_{2g}$

[a]) $Dq = 2050\ cm^{-1}$, $B = 410\ cm^{-1}$, $\zeta = 1000\ cm^{-1}$, $C/B = 4.90$.

d) Hexafluoropalladate (IV), PdF_6^{2-}

Potassium hexafluoropalladate(IV), K_2PdF_6, may be obtained by the action of BrF_3 on K_2PdCl_4 (*35*) and the Rb and Cs salts by direct fluorination of the corresponding $PdCl_{4}^{2-}$ compounds (*36*). The potassium salt was found to be diamagnetic so that the ground state is the expected $^1A_{1g}$ (t_{2g}^6) low-spin state. The electronic spectrum was studied by diffuse reflectance by *Brown et al.* (*32*) and also briefly reported by *Allen et al.* (*12*). Two well marked peaks of moderate intensity were found at 25.0 and 30.0 kK., and identified as the $^1A_{1g} \rightarrow {}^1T_{1g}$ and $^1A_{1g} \rightarrow {}^1T_{2g}$ *d—d* transitions respectively, whilst a weaker broad shoulder at 21.0 kK. was assigned as the spin-forbidden excitation, $^1A_{1g} \rightarrow {}^3T_{1g}$. This latter band was however too close to the stronger peak at 25.0 kK. for any meaningful calculations of relative intensities [c.f. NiF_6^{2-} (*37*)] to be made. Calculations showed that the other expected spin-forbidden band, $^1A_{1g} \rightarrow {}^3T_{2g}$, (which also corresponds to a one-electron excitation) would be obscured by the 25 kK. absorption, and the data are satisfactorily fitted with $Dq = 2600\ cm^{-1}$ and $B = 400\ cm^{-1}$, using $C/B = 5.0$.

No bands were reported by *Brown et al.* corresponding to charge-transfer transitions, and their absence below 50 kK. was confirmed by *Allen et al.* Since the lowest Laporte-allowed band for the $PdCl_6^{2-}$ anion is found close to 30 kK. (*24*), this is as would be expected from optical electronegativity considerations.

Table 8. Spectroscopic data for caesium hexafluoropalladate (IV)

Band Position (kK.)	Assignment
21.0 (br)	${}^1A_{1g} \rightarrow {}^3T_{1g}$
25.0	${}^1A_{1g} \rightarrow {}^1T_{1g}$
30.0	${}^1A_{1g} \rightarrow {}^1T_{2g}$

$Dq = 2600$ cm^{-1}, $B = 400$ cm^{-1}, $C/B = 5.0$.

(iii) MF_6^- Systems

For the 4*d* series the M(V) oxidation state is found for Mo, Tc, and Ru as hexafluoro anions. These MF_6^- complexes are all quite strongly oxidising, and their stability again decreases towards the end of the series. Some spectroscopic data are available for all three systems and are reviewed below.

a) *Hexafluoromolybdate* (V), MoF_6^-

The K, Rb, and Cs salts of this anion are prepared by the action of the appropriate alkali metal iodide on MoF_6 in IF_5 (*38*), and potassium hexafluoromolybdate (V), $KMoF_6$, is found to have a magnetic moment of 1.3 B.M. at 298 °K (*39*). The electronic spectrum of $CsMoF_6$ has been reported by *Brown et al.* as showing bands at 24, 29, and 35 kK., and the first of these was assumed to correspond to the single expected *d—d* band, ${}^2T_{2g} \rightarrow {}^2E_g$. The remaining bands were attributed to spin-orbit splittings or charge-transfer absorptions, but neither of these explanations appears at all likely. Thus for the analogous $MoCl_6^-$ anion the lowest Laporte-allowed band is reported at 30.2 kK. (*40*), so that for MoF_6^- the $\pi \rightarrow t_{2g}$ bands would not be expected below 50 kK. Similarly for Mo (V) ξ is only of the order of 900 cm^{-1} so that splittings of as much as 5 kK. are out of the question. Lacking diagrammatic presentation it is probably imprudent even to estimate *Dq* and a further examination of this spectrum is clearly needed.

b) *Hexafluorotechnetate* (V), TcF_6^-

The Na, K, Rb, and Cs salts of this anion are prepared by the interaction of the appropriate alkali metal chloride with a slight excess of TcF_6 in IF_5 (*41*). For Na, Rb, and Cs hexafluorotechnetate (V) magnetic moments ranging between 1.75 and 2.25 B.M. between 90° and 295 °K were found with slightly larger values for the potassium salt. The electronic spectrum of $CsTcF_6$ was measured by diffuse reflectance and absorptions found at 13.6, 17.8, 23.0, and 32.0 kK., of which the last constituted the most prominent band and was assigned as the spin-allowed *d—d* transition, $^3T_{1g} \rightarrow {}^3T_{2g}$. The 17.8 kK. peak was noted as being sharp and narrow and was therefore assigned as the spin-forbidden $^3T_{1g}$ $(t_{2g}^2) \rightarrow {}^1A_{1g}$ (t_{2g}^2) *d—d* transition, but the remaining bands were attributed either to impurities or to unspecified symmetry or spin-orbit effects. With these assumptions the parameters $Dq = 3460$ cm^{-1} and $B = 520$ cm^{-1} were derived, using $C/B = 4.4$.

This assignment though is disquieting on a number of grounds. In the first place the derived value for 10 Dq is very much larger (6 kK.) than any other found for a 4*d* hexafluoro complex, and over 8 kK. greater than that for RuF_6^- (q.v.). Secondly the band at 13.6 kK. is much too high in energy to represent the only other plausible *d—d* transition, $^3T_{1g} \rightarrow {}^1E_g$, $^1T_{2g}$, even allowing for spin-orbit coupling and lower symmetry effects, whilst if the 17.8 kK. band *is* $^3T_{1g} \rightarrow {}^1A_{1g}$ it cannot be split by either of these contributions. Finally no *d—d* assignment seems possible at all for the 23.0 kK. band.

However, an alternative approach is possible since the 32.0 kK. band was described by *Hugill* and *Peacock* as being very strong thus suggesting that a charge-transfer assignment is a possibility. The 23.0 kK. peak could then be attributed to the $^3T_{1g} \rightarrow {}^3T_{2g}$ transition and the narrow peak at 17.8 kK. as $^3T_{1g} \rightarrow {}^1A_{1g}$ as before, thus leaving only the 13.6 kK.

Table 9. Spectroscopic data for potassium hexafluorotechnetate (V)

Band Position (kK.)	Assignment
13.6	?
17.8	$^3T_{1g} \rightarrow {}^1A_{1g}$
23.0	a) or $^3T_{1g} \rightarrow {}^3T_{2g}$ b)
32.0	$^3T_{1g} \rightarrow {}^3T_{2g}$ a) or $\pi \rightarrow t_{2g}$ b)

a) Ref. (*41*): $Dq = 3460$ cm^{-1}, $B = 520$ cm^{-1}.
b) Present work: $Dq = 2500$ cm^{-1}, $B = 535$ cm^{-1}.

absorption to be ascribed to possible impurities. Taking $C/B = 4.5$ this yields the parameters $Dq = 2500$ cm^{-1} and $B = 540$ cm^{-1} (c.f. $Dq =$) 2600 cm^{-1} and $B = 425$ cm^{-1} for RuF_6^-). Lacking diagrammatic presentation of the spectrum or details of relative intensities this affords as reasonable an interpretation as that originally suggested, and leads to a $\varkappa_{opt}$ value for Tc(V) of 2.60 which is quite a feasible result on the basis of the 2.80 found for Ru(V) and the 2.25 for Tc(IV). (See also Section 5).

c) *Hexafluororuthenate* (V), RuF_6^-

Potassium hexafluororuthenate(V), $KRuF_6$, is obtained by the action of BrF_3 on a 1:1 mixture of ruthenium and KCl, according to *Hepworth, Peacock,* and *Robinson* (30), and a magnetic moment of between 3.6 and 3.8 B.M. at 298 °K was found. Again both *Brown et al.* and *Allen et al.* have studied the diffuse reflectance spectrum of the RuF_6^- ion, the former workers reporting weak bands at 10.4 and 15.4 kK., and stronger absorptions at 24.2, 25.2, and 33.3 kK. The findings of *Allen et al.* are generally in agreement with these results; they reported three peaks in the lower energy region, at 9.2, 10.1, and 15.7 kK., all of which were found, contrary to the results of *Brown et al.*, to be of relatively low intensity, and which were therefore assigned as the spin-forbidden transitions $^4A_{2g} \rightarrow {}^2E_g$, $^4A_{2g} \rightarrow {}^2T_{1g}$, and $^4A_{2g} \rightarrow {}^2T_{2g}$ respectively. At higher energies the stronger band at 26.4 kK. and a shoulder at 32.0 kK. were assigned as the spin-allowed *d—d* excitations, $^4A_{2g} \rightarrow {}^4T_{2g}$, and $^4A_{2g} \rightarrow$

Table 10. Spectroscopic data for potassium hexafluororuthenate (V)

Band Position (kK.) Obsd.	Calcd.[a])	Assignment
9.2	9.4 (Γ_8)	$^4A_{2g} \rightarrow {}^2E_g$
10.1	9.6 (Γ_8)	$^4A_{2g} \rightarrow {}^2T_{1g}$
	9.9 (Γ_6)	
15.7	15.4 (Γ_8)	$^4A_{2g} \rightarrow {}^2T_{2g}$
	15.5 (Γ_7)	
(22.0)	—	?
26.4	26.1	$^4A_{2g} \rightarrow {}^4T_{2g}$
32.0 (sh)	31.0	$^4A_{2g} \rightarrow {}^4T_{1g}$
40.0	—	$\pi \rightarrow t_{2g}$
50.0	—	$\pi \rightarrow t_{2g}$ or $\sigma \rightarrow t_{2g}$

[a]) $Dq = 2600$ cm^{-1}, $B = 425$ cm^{-1}, $\zeta = 1000$ cm^{-1}, $C/B = 4.75$.

$^4T_{1g}$, but the traditional ligand field approach afforded no explanation for the broad absorption at 22.0 kK. (c.f. the 24.2 kK. band of *Brown et al.*) which was ascribed either to impurities or to a single-photon two-ion electronic excitation. Fitting of the bands using $\zeta = 1000$ cm^{-1} and $C/B = 4.75$ afforded the parameters $Dq = 2600$ cm^{-1} and $B = 425$ cm^{-1}. Note here that for the d^3 $^4A_{2g}$ (t_{2g}^3) systems no spin-orbit splitting of the ground state is to be expected since in O* $^4A_{2g}$ transforms as Γ_8 and is therefore not split.

Above the d–d bands more intense absorptions were found at 40.0 and near 50 kK., (Fig. 2). The former of these is clearly a Laporte-allowed charge-transfer ($\pi \rightarrow t_{2g}$) band, but for the 50 kK. band either the $\pi \rightarrow t_{2g}$ or a $\sigma \rightarrow t_{2g}$ assignment is possible.

(iv) MF_6 Systems

The M(VI) oxidation state is represented in the 4*d* series by the hexafluorides, MF_6, of the elements Mo, Tc, Ru, and Rh. All are obtained by direct fluorination of the metal and are unstable powerfully oxidising species — once again the instability seems most marked at the end of the series. Unfortunately hardly any electronic spectral data exist. The first charge-transfer band of the d^0 MoF_6 has been located at 54 kK. (*42*), and a study of the vibrational spectrum of RuF_6 (*43*) revealed electronic bands at 1.95 and 1.4 kK., which are probably the Γ_3, $\Gamma_5 \rightarrow \Gamma_1$, and Γ_3, $\Gamma_5 \rightarrow \Gamma_4$ transitions within the spin-orbit split $^3T_{1g}$ (t_{2g}^2) ground state. Apart however from these observations no results are available.

3. Electronic Spectra of the Hexafluoro Species of the 5*d* Series

(i) MF_6^{2-} Systems

In the 5*d* series the stability of the higher oxidation state fluoro species becomes pronounced. Thus, with the possible exception of Ir, for which the preparation of a diamagnetic K_3IrF_6 complex has been claimed (*44*), no MF_6^{3-} complexes are known, whilst the M(IV) oxidation state is represented by the MF_6^{2-} anions of Re, Os, Ir, and Pt. These latter compounds though show no significant oxidising tendencies and are all quite stable in aqueous solution.

As noted in Section 1.(i) the values of Dq are again larger in general in the 5*d* series, relative to the 4*d* values, whilst smaller B values are usually observed. The uniform low-spin behaviour for d^4, d^5, and d^6

systems is therefore found in the third transition series also. Satisfactory spectroscopic data are available for all four MF_6^{2-} species, and in some cases it is possible to deduce values of ζ, the effective spin-orbit coupling constant, as well as those for Dq and B. Trends in both B and ζ, as reflected in the nephelauxetic and relativistic ratios, β and β^* respectively, are discussed in Section 4.

a) *Hexafluororhenate* (IV), ReF_6^{2-}

Potassium hexafluororhenate (IV), K_2ReF_6, is, obtained from potassium perrhenate via $K_2Re(I)_6$ and subsequent fusion with KHF_2 (*45*), and shows a magnetic moment of 3.3—3.4 B.M. at 298 °K (*31, 46*). The electronic spectrum in aqueous solution has been studies in some detail by *Jørgensen* and *Schwochau* (*29*), and the principle features of their results are listed in Table 11. (See also Fig. 6).

At lower energies the rather weak band at 9—11 kK., corresponding to the spin-forbidden transitions $^4A_{2g} \rightarrow {}^2E_g$ and $^4A_{2g}, \rightarrow {}^2T_{1g}$, shows considerable vibrational structure, as well as splittings due to the considerable spin-orbit interaction, thereby complicating the assignments somewhat. In O* ^2E_g becomes Γ_8 whilst $^2T_{1g}$ yields $\Gamma_6 + \Gamma_8$, and we suggest the assignments $^4A_{2g} \rightarrow a\Gamma_8$, $^4A_{2g} \rightarrow b\Gamma_8$, and $^4A_{2g} \rightarrow {}^2T_{1g}$ (Γ_6), for the 9.1, 10.1, and 10.9 kK. maxima respectively. Note that the two Γ_8 components will be extensively mixed so that the descriptions ^2E_g or $^2T_{1g}$ will no longer be really appropriate. The weak band between 17 and 19 kK. may also clearly be attributed to a spin-forbidden *d—d* transition, here $^4A_{2g} \rightarrow {}^2T_{2g}$. Again extensive vibrational structure was observed but the main peaks at 17.7 and 19.0 kK. may reasonably be ascribed to the Γ_7 and Γ_8 components respectively of $^2T_{2g}$.

The spin-forbidden transitions all appear as sharp narrow absorptions since they correspond to intra-subshell transitions within the t_{2g}^3 manifold. Using $C/B = 4$ *Jørgensen* and *Schwochau* (*29*) obtained the parameters $Dq = 3280$ cm^{-1}, $B_{55} = 543$ cm^{-1}, and $\zeta = 2550$ cm^{-1} ($\pm 20\%$), but only one excited quartet level could be identified with certainty. Thus the moderately strong band at 32.8 kK. was assigned as $^4A_{2g} \rightarrow {}^4T_{2g}$, but the less well defined maximum at 37.5 kK. could only tentatively be ascribed to the $^4A_{2g} \rightarrow {}^4T_{1g}$ excitation since either this or the broad absorption at around 41.7 kK. could have been due to traces of ReO_4^-. The assignment of the moderately strong band at 48.3 kK. is also problematical: possibly it represents a $\pi_g \rightarrow t_{2g}$ charge-transfer band but the diffuse reflectance spectrum of *Allen et al.* (*11*) did not give any indication of charge-transfer transitions below 50 kK.

It may here be noted that although the Re(IV) spin-orbit coupling constant is easily large enough to cause observable splittings of the narrow quartet → doublet bands, it is not sufficient to split the spin-allowed peaks. For both the ${}^4T_{2g}$ and ${}^4T_{1g}$ levels the overall first order splitting is only 2/3 ζ, *i.e.* ~1700 cm^{-1}, but with band widths for the $t_{2g}^3 \rightarrow t_{2g}^2\, e_g$ excitations of the order of 2kK. or more it is unlikely that such effects would be detectable.

Table 11. Spectroscopic data for potassium hexafluororhenate (IV)

Band Position (kK.)	Assignment
9.1	${}^4A_{2g} \rightarrow {}^2E_g\ (\Gamma_8)$
10.1	${}^4A_{2g} \rightarrow {}^2T_{1g}\ (\Gamma_8)$
10.9	${}^4A_{2g} \rightarrow {}^2T_{1g}\ (\Gamma_6)$
17.7	${}^4A_{2g} \rightarrow {}^2T_{2g}\ (\Gamma_7)$
19.0	${}^4A_{2g} \rightarrow {}^2T_{2g}\ (\Gamma_8)$
32.8	${}^4A_{2g} \rightarrow {}^4T_{2g}$
(37.5)	${}^4A_{2g} \rightarrow {}^4T_{1g}$?
(41.7)	ReO_4^- ?
(48.3)	$\pi_g \rightarrow t_{2g}$?

$Dq = 3280$ cm^{-1}, $B_{55} = 543$ cm^{-1}, $\zeta = 2550$ cm^{-1}.

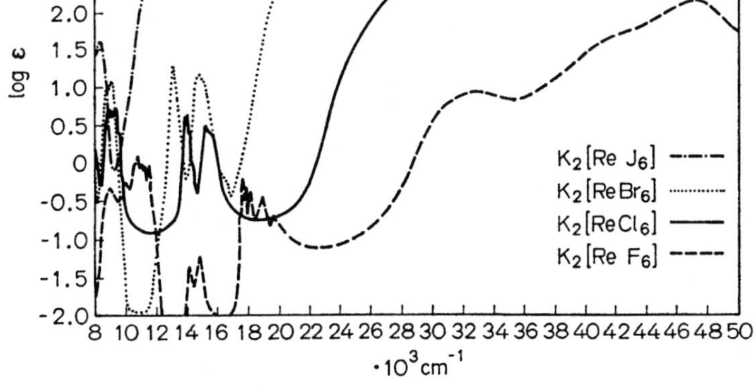

Fig. 6. Electronic spectrum of K_2ReF_6

b) Hexafluoroosmate (IV), OsF_6^{2-}

Salts of the hexafluoroosmate (IV) anion, OsF_6^{2-}, are readily obtained by the hydrolysis of the corresponding hexafluoroosmate(V) species (q.v.) (*47, 48*), and for K_2OsF_6 a magnetic moment of 1.48 B.M. at 300 °K has been reported (*49*), which is in accordance with the expectation of a low-spin $^3T_{1g}$ (t_{2g}^4) ground state. Some early spectroscopic measurements were made for OsF_6^{2-} in aqueous solution by *Hepworth et al.* (*48*), who reported a weak band at 32.5 kK. and a much stronger one just above 50 kK., but more recently diffuse reflectance studies have been reported by *Brown et al.* (*32*) and by *Allen et al.* (*11*).

For Cs_2OsF_6 *Brown et al.* recorded only three bands — at 23.5, 30.0, and 33.0 kK. — in the region between 10 and 40 kK., the lowest energy absorption being attributed to a $^3T_{1g} \rightarrow {}^5E_g$ transition, and the other two bands to spin-allowed, $t_{2g}^4 \rightarrow t_{2g}^3\ e_g$ *d—d* excitations. However, using K_2OsF_6, *Allen et al.* obtained significantly different results (Fig. 7). Thus two fairly weak bands were detected at 12.7 and 18.5 kK., with a shoulder at 22.5 kK. and stronger absorptions at about 30 and 42 kK., the former being very broad. The start of an intense band above 50 kK. was also evident (c.f. *Hepworth et al.*) and below 10 kK. a broad moderately strong band was found at 5.6 kK. together with evidence of a further peak below 4 kK.

The band at 12.7 kK. was readily assigned as $^3T_{1g} \rightarrow {}^1T_{2g}, {}^1E_g$, and by comparison with the data of *Dorain et al.* (*50*) and of *Dickinson* and *Johnson* (*51*) for the related $OsCl_{6^{2-}}$ ion, the bands at 18.5 and 22.5 kK. were assigned as $^3T_{1g} \rightarrow {}^5E_g$ and $^3T_{1g} \rightarrow {}^1A_{1g}$ respectively. All these bands therefore formally correspond to spin-forbidden transitions but because of the very substantial values of ζ found in the 5*d* series their relative intensities are appreciably higher than those encountered in the first and second transition series. Assuming that the 23.5 kK. band of *Brown et al.* corresponds to the 22.5 kK. absorption of *Allen, El-Sharkawy,* and *Warren,* there is a disagreement over its assignment. However *Brown et al.* reached their conclusion without consideration of the large spin-orbit effects and *Allen et al.* showed that when this was included in the calculation of the fitting parameters all the observed levels could be adequately reproduced (vide infra).

In the low energy region the measurements of *Allen et al.* went down only to 4 kK. but hexachlorobutadiene mull data indicated the presence of a further band at 3.0—3.4 kK.. This and the 5.6 kK. band were therefore readily identified as transitions within the spin-orbit split ground state manifold — the 3 kK. absorption as $^3T_{1g}$ (Γ_1) $\rightarrow {}^3T_{1g}$ (Γ_4) and the 5.6 kK. peak as $^3T_{1g}$ (Γ_1) $\rightarrow {}^3T_{1g}$ (Γ_3, Γ_5). *Allen et al.* were unable to resolve the broad band at 30 kK. but it may be noted that six formally spin-allowed $^3T_{1g}$ $(t_{2g}^4) \rightarrow t_{2g}^3\, e_g$ transitions would be anticipated in this region, viz: — $^3T_{1g} \rightarrow {}^3T_{1g}$, ^3E_g, $^3T_{2g}$, $^3A_{2g}$, $^3A_{1g}$, and ^3E_g. The 42.0 kK. peak was also assigned to *d–d* transitions to higher lying $^3T_{1g}$ and $^3T_{2g}$ levels, but the intense peak above 50 kK. was attributed to a Laporte-allowed $\pi \rightarrow t_{2g}$ transition on the basis of the similar bands for the related $OsCl_6^{2-}$ being located at about 25 kK. (*24*). Because of the large splitting of the $^3T_{1g}$ ground state the effective ζ value was determined

Table 12. Spectroscopic data for potassium hexafluoroosmate(IV)

Band Position (kK.)		Assignment
Obsd.	Calcd.[a]	
3.2	3.1	$^3T_{1g}(\Gamma_1) \rightarrow {}^3T_{1g}(\Gamma_4)$
5.6	5.5	$^3T_{1g}(\Gamma_1) \rightarrow {}^3T_{1g}(\Gamma_3, \Gamma_5)$
12.7	12.5, 12.95	$^3T_{1g}(\Gamma_1) \rightarrow {}^1T_{2g}, {}^1E_g$
18.5	17.3	$^3T_{1g}(\Gamma_1) \rightarrow {}^5E_g$
22.5	21.5	$^3T_{1g}(\Gamma_1) \rightarrow {}^1A_{1g}$
30.0 (br)	28.9, 29.6, 30.3, 30.6	$^3T_{1g}(\Gamma_1) \rightarrow {}^3E_g, {}^3T_{1g}, {}^3T_{2g}, {}^3A_{2g}$
	31.3, 33.3	$^3A_{1g}, {}^3E_g\ (t_{2g}^3\, e_g)$
42.0	41.0, 42.5	$^3T_{1g}(\Gamma_1) \rightarrow {}^3T_{1g}, {}^3T_{2g}\ (t_{2g}^3\, e_g)$
~50	—	$(\pi + \sigma)\ t_{1u} \rightarrow t_{2g}$

[a]) $Dq = 2600$ cm^{-1}, $B = 500$ cm^{-1}, $\zeta = 2900$ cm^{-1}, $C/B = 4.75$.

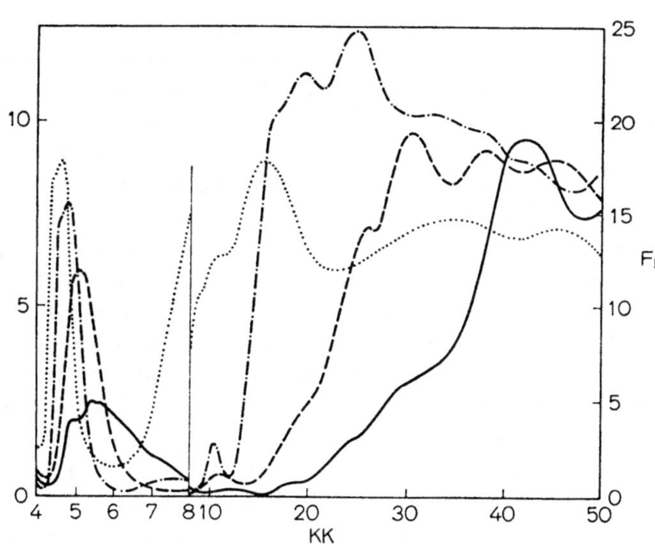

Fig. 7. Electronic spectrum of K_2OsF_6 ———, K_2OsCl_6 - - - - - -, K_2OsBr_6 -.-.-.-.-, and K_2OsI_6 ....

with some accuracy (the first order estimate of 3/2 ζ is actually a slight underestimate), and the data were well fitted by the parameters $Dq = 2600\ cm^{-1}$, $B = 500\ cm^{-1}$, and $\zeta = 2900\ cm^{-1}$, with $C/B = 4.75$.

c) *Hexafluoroiridate*(IV), IrF_6^{2-}

Hexafluoroiridate(IV) salts are again obtained by hydrolysis of the corresponding M(V) species, MF_6^- (*48*). For caesium hexafluoroiridate (IV) Cs_2IrF_6, a magnetic moment of 1.42 B.M. at 298 °K has been reported (*48*), confirming that the ground state is the low-spin $^2T_{2g}$ (t_{2g}^5). For the hexafluoroiridate(IV) anion *Hepworth et al.* (*48*) reported two bands in the solution spectrum — an intense absorption at 46.9 kK. and a much weaker band at 31.6 kK. — but *Brown et al.* and *Allen et al.* have both carried out diffuse reflectance measurements.

For Cs_2IrF_6 *Brown et al.* (*32*) found peaks at 13.0, 19.0, 24.2, 30.5, and 34.0 kK., and these were assigned as transitions from the $^2T_{2g}$ ground state to the $^4T_{1g}$, $^4T_{2g}$, $^2A_{2g}$, and $^2T_{1g}$, ^2E_g, and $^2A_{1g}$ levels respectively. The 19.0 kK. peak was though said to be very strong and an alternative charge-transfer assignment was considered possible. This latter possibility must however be discounted since in the related $IrCl_6^{2-}$ anion the lowest Laporte-allowed $\pi \rightarrow t_{2g}$ transition is found at 20.0 kK., so that in IrF_6^{2-} the corresponding bands would not be expected below about 45 kK.

However, for both K_2IrF_6 and Cs_2IrF_6 *Allen et al.* (*11*) found only a very faint absorption at 13 kK., which they thought to be spurious. (Fig. 8). A band of quite normal intensity at 19.8 kK. was assigned as $^2T_{2g} \rightarrow {}^4T_{1g}$ and a further band of comparable intensity at 24.9 kK. was similarly ascribed to the other expected spin-forbidden transition, $^2T_{2g} \rightarrow {}^4T_{2g}$. The former assignment was supported by the observation of the $^2T_{2g} \rightarrow {}^4T_{1g}$ band between 19 and 21 kK. in $IrCl_6^{2-}$ (*52*), and it was shown by *Allen et al.* that no great change in the energy of this transition would be expected between $IrCl_6^{2-}$ and IrF_6^{2-}. Furthermore, the calculations of *Allen et al.* showed that the $^4T_{1g}$ level should be split by some 4.4 kK. due to spin-orbit coupling, whilst the $^4T_{2g}$ level should show only a splitting of about 0.6 kK.. Thus the observation of a broad band at 19.8 kK. and of a fairly narrow one at 24.9 kK. supported the listed assignment. Also of course, although formally spin-forbidden, these bands are quite strong relative to the spin-allowed *d—d* bands because of the substantial spin-orbit interaction in the 5*d* series.

At higher energies *Allen et al.* recorded stronger broad absorptions at 30.0 and 38.0 kK. attributed to the spin-allowed $t_{2g}^5 \rightarrow t_{2g}^4 e_g$ transitions $^2T_{2g} \rightarrow {}^2T_{1g}$, $^2T_{2g}$, $^2A_{2g}$, ^2E_g and $^2T_{2g} \rightarrow {}^2T_{2g}$, $^2T_{1g}$, $^2A_{1g}$ respectively, and an intense band at 47.8 kK. attributed to a *Laporte*-allowed $\pi \rightarrow t_{2g}$ band. In the low energy region a moderately strong band at 6.65 kK. was once more ascribed to a transition within the spin-orbit split $^2T_{2g}$ ground state, *i.e.* $^2T_{2g}$ $(\Gamma_7) \rightarrow {}^2T_{2g}$ (Γ_8). To the first order the splitting of the ground level should be 3/2 ζ, but a complete strong field spin-orbit calculation indicated this to be a slight underestimate and afforded the parameters $Dq = 2700$ cm^{-1}, $B = 510$ cm^{-1}, and $\zeta = 3300$ cm^{-1} with $C/B = 4.90$.

Table 13. Spectroscopic data for potassium hexafluoroiridate(IV)

Band Position (kK.)		Assignment
Obsd.	Calcd.[a]	
6.65	6.3	$^2T_{2g}$ $(\Gamma_7) \rightarrow {}^2T_{2g}$ (Γ_8)
19.8	17.3, 19.2, 20.6, 21.7	$^2T_{2g}$ $(\Gamma_7) \rightarrow {}^4T_{1g}$ $(\Gamma_8, \Gamma_7, \Gamma_8, \Gamma_6)$
24.9	24.2, 24.55, 24.8, 24.8	$^2T_{2g}$ $(\Gamma_7) \rightarrow {}^4T_{2g}$ $(\Gamma_6, \Gamma_8, \Gamma_7, \Gamma_8)$
30.0 (br)	28.5, 31.3, 31.6, 33.3	$^2T_{2g}$ $(\Gamma_7) \rightarrow {}^2T_{1g}, {}^2T_{2g}, {}^2A_{2g}, {}^2E_g,$
38.0 (br)	35.9, 37.3, 39.6	$^2T_{2g}, {}^2T_{1g}, {}^2A_{1g}, (t_{2g}^4\, e_g)$
47.8	—	$(\pi + \sigma)\ t_{1u} \rightarrow t_{2g}$

[a]) $Dq = 2700$ cm^{-1}, $B = 510$ cm^{-1}, $\zeta = 3300$ cm^{-1}, $C/B = 4.90$.

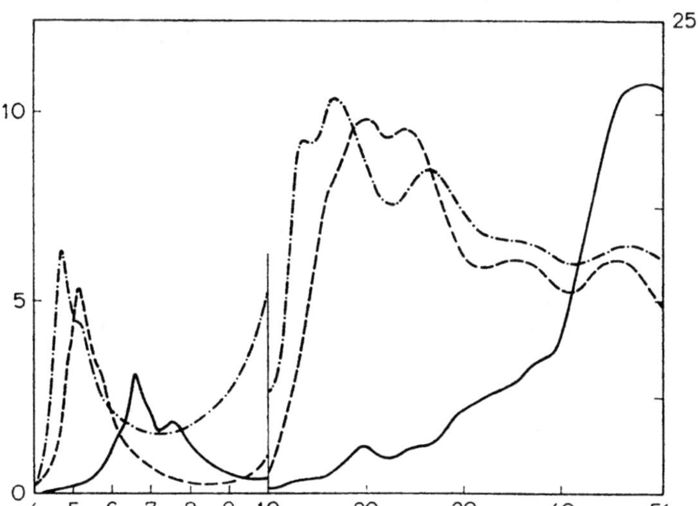

Fig. 8. Electronic spectrum of K_2IrF_6 ———, K_2IrCl_6 - - - - - -, and K_2IrBr_6 -·-·-·-·

d) *Hexafluoroplatinate*(IV), PtF_6^{2-}

Hexafluoroplatinate(IV) salts may conveniently be obtained by the action of either F_2 or BrF_3 on the appropriate hexachloroplatinate(IV) salts. The spectrum of the PtF_6^{2-} anion in aqueous solution has been reported by *Wheeler, Perros*, and *Naeser* (*53*) and an amplified account has been quoted by *Jørgensen* (*54*). Bands of moderate intensity were found at 31.4 and 36.4 kK. ($\varepsilon = 33$ and 24 respectively), and two weaker absorptions were detected at 22.5 and 24.5 kK. ($\varepsilon = 4$ and 5). On the assumption of a low-spin $^1A_{1g}$ (t_{2g}^6) ground state the latter two bands may be assigned as the spin-forbidden *d–d* transitions, $^1A_{1g} \rightarrow {}^3T_{1g}$ and $^1A_{1g} \rightarrow {}^3T_{2g}$, and the 31.4 and 36.4 kK. peaks as the spin-allowed excitations $^1A_{1g} \rightarrow {}^1T_{1g}$ and $^1A_{1g} \rightarrow {}^1T_{2g}$. The spectrum of Cs_2PtF_6 has also been studied by diffuse reflectance by *Brown et al.*, very similar results being obtained. Thus a broad band assigned as $^1A_{1g} \rightarrow {}^3T_{1g}$, $^3T_{2g}$ was observed between 21 and 28 kK. together with peaks at 32.5 and 37.0 kK., ascribed as above as $^1A_{1g} \rightarrow {}^1T_{1g}$ and $^1A_{1g} \rightarrow {}^1T_{2g}$, the data yielding the parameters $Dq = 3300$ cm^{-1} and $B = 380$ cm^{-1}.

The spin-forbidden bands, $^1A_{1g} \rightarrow {}^3T_{1g}$ and $^1A_{1g} \rightarrow {}^3T_{2g}$, both derive most of their intensity via spin-orbit interaction with the lowest spin-allowed level, $^1T_{1g}$. The Γ_4 matrix elements $\langle {}^3T_{1g} \| {}^1T_{1g} \rangle$ and $\langle {}^3T_{2g} \| {}^1T_{1g} \rangle$ are respectively $(4)^{-1}\,(2)^{1/2}\,\zeta$ and $(4)^{-1}\,(6)^{1/2}\,\zeta$, so that perturbation calculations suggest an intensity relative to $^1A_{1g} \rightarrow {}^1T_{1g}$ of ca. $(40)^{-1}$ for $^3T_{1g}$ and ca. $(8)^{-1}$ for $^3T_{2g}$, taking ζ Pt(IV) as 4000 cm^{-1}. Lacking band width data these results cannot be checked directly against the experimental values but are clearly of a reasonable order of magnitude.

As expected from the observation of $\pi \rightarrow e_g$ for $PtCl_6^{2-}$ at 38.2 kK. (*24*), no charge-transfer bands were observed by either set of workers.

Table 14. Spectroscopic data for potassium hexafluoroplatinate(IV)

Band Position (kK.)	Assignment
22.5	$^1A_{1g} \rightarrow {}^3T_{1g}$
24.5	$^1A_{1g} \rightarrow {}^3T_{2g}$
31.4	$^1A_{1g} \rightarrow {}^1T_{1g}$
36.4	$^1A_{1g} \rightarrow {}^1T_{2g}$

$Dq = 3300$ cm^{-1}, $B = 380$ cm^{-1}.

(ii) MF_6^- Systems

In the 5*d* series the M(V) oxidation state is the most widely occurring condition as far as hexafluoro salts are concerned. The MF_6^- anion is thus known for the elements from W to Au, but unlike the M(IV) complexes these are all strongly oxidising species; they are unstable towards water and the MF_6^- ions of Re, Os, Ir, and Pt all yield the corresponding quadrivalent salts on hydrolysis.

Detailed spectroscopic results have been recorded (*10*, *32*) for the OsF_6^- and IrF_6^- anions, and some data are available for the WF_6^- and ReF_6^- and PtF_6^- species. No spectrum has though been obtained for the recently synthesised $CsAuF_6$ [prepared from AuF_3 and XeF_2 with subsequent reaction with KF (*55*)].

a) *Hexafluorotungstate*(V), WF_6^-

Caesium hexafluorotungstate(V), $CsWF_6$, is obtained (*37*) by the action of CsI on WF_6 in liquid sulphur dioxide, and has been found (*38*) to have a magnetic moment of ca. 0.50 B.M. at 298 °K, consistent with a $^2T_{2g}$ (t_{2g}^1) ground state ($\zeta \gg$kT). *Brown et al.* (*32*) by diffuse reflectance observed a single peak at 32.4 kK. which they assigned as the *d—d* excitation $^2T_{2g} \rightarrow {}^2E_g$ (*i.e.* $Dq = 3240$ cm^{-1}). No charge-transfer bands were found but this is not surprising since the lowest $\pi_u \rightarrow t_{2g}$ band for the related WCl_6^- anion is found at 29.4 kK. (*56*).

b) *Hexafluororhenate*(V), ReF_6^-

Hexafluororhenate(V) salts are obtained by the interaction of the appropriate alkali metal iodide with ReF_6 in either SO_2 or IF_5 (*57*). For most cations the magnetic moments lie between about 1.3 and 1.4 B.M. at 298 °K, slightly higher results being found for the potassium salt (*39*).

The diffuse reflectance spectrum of $CsReF_6$ was studied between 10 and 40 kK. by *Brown et al.* (*32*), but no clearly defined peaks could be recognised above a high background absorption. Since Laporte-allowed charge-transfer bands usually yield well marked intense absorptions it is reasonable therefore to conclude that these are absent below 40 kK. (See also Section 5).

In any future reexamination of the spectrum of the ReF_6^- anion it may be anticipated that a moderately strong band should be found between ca. 5–10 kK., since the spin-orbit splitting of the $^3T_{1g}$ ground level again amounts to 3/2 ζ in the first order.

c) *Hexafluoroosmate*(V), OsF_6^-

The hexafluoroosmate(V) salts are readily obtained by the action of BrF_3 on 1:1 mixtures of $OsBr_4$ and the appropriate alkali metal chloride

(*47*, *48*). A magnetic moment of 3.23 B.M. at 298 °K was found for $CsOsF_6$ (*46*), and diffuse reflectance studies of the spectrum have been reported by *Brown et al.* (*32*) and by *Allen et al.* (*10*).

The former authors recorded bands at 10.4, 11.0, 18.0, 31.0, and 39.0 kK. which they assigned as transitions from a $^4A_{2g}$ ground state to the ^2E_g, $^2T_{1g}$, $^2T_{2g}$, $^4T_{1g}$, and $^4T_{2g}$ levels respectively. The spectrum of *Allen et al.* (Fig. 9) however covered a more extended range and a number of additional features were noted. Thus the rather weak absorptions found at 8.0, 10.3, and 10.8 kK. were attributed to the $^4A_{2g} \rightarrow {}^2E_g$ and $^4A_{2g} \rightarrow {}^2T_{1g}$ excitations: the ^2E_g (Γ_8) and $^2T_{1g}$ (Γ_8) states are extensively mixed but inspection of the eigenvectors from diagonalisation of the energy matrices indicates that the preponderant configurations are as shown in Table 15, whilst the 10.8 kK. peak corresponds to the $^2T_{1g}$ (Γ_6) component.

At 16.3 and 18.2 kK. *Allen et al.* found two fairly sharp but rather weak peaks instead of the single absorption reported at 18 kK. by *Brown et al.* These were readily attributed to the Γ_7 and Γ_8 components respectively of the $^4A_{2g} \rightarrow {}^2T_{2g}$ transition, and since their separation is strongly dependent on ζ, a good estimate of the effective spin-orbit coupling constant was thereby derived. At higher energies a broad very intense band was found beginning at about 40 kK. with a maximum at 37.5 kK. and a shoulder at 41.7 kK. These absorptions could conceivably have corresponded to the $^4A_{2g} \rightarrow {}^4T_{2g}$ and $^4A_{2g} \rightarrow {}^4T_{1g}$ transitions, but were considered improbably strong for a *d–d* band and were assigned on optical electronegativity considerations as belonging to a Laporte-allowed $\pi \rightarrow t_{2g}$ band. Despite a careful search no evidence of a distinct absorption at 31 kK. was found, and the *d–d* bands were well fitted by the parameters $Dq = 3500$ cm^{-1}, $B = 410$ cm^{-1}, $\zeta = 3200$ cm^{-1}, with $C/B = 4.75$.

Table 15. Spectroscopic data for caesium hexafluoroosmate(V)

Band Position (kK.)	Assignment
8.0	$^4A_{2g} \rightarrow {}^2T_{1g}$ (Γ_8)
10.3	$^4A_{2g} \rightarrow {}^2E_g$ (Γ_8)
10.8	$^4A_{2g} \rightarrow {}^2T_{1g}$ (Γ_6)
16.3	$^4A_{2g} \rightarrow {}^2T_{2g}$ (Γ_7)
18.2	$^4A_{2g} \rightarrow {}^2T_{2g}$ (Γ_8)
37.5	$^4A_{2g} \rightarrow {}^4T_{2g}$ ( ) or $\pi \rightarrow t_{2g}$
41.7 (sh)	$\pi \rightarrow t_{2g}$

$Dq = 3500$ cm^{-1}, $B = 410$ cm^{-1}, $\zeta = 3200$ cm^{-1}, $C/B = 4.75$.

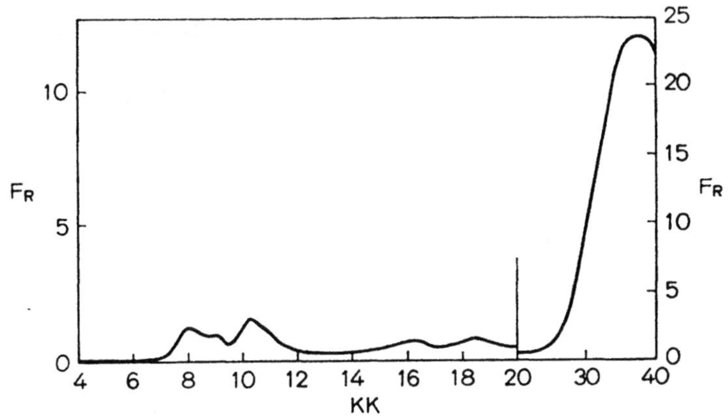

Fig. 9. Electronic spectrum of $CsOsF_6$

d) Hexafluoroiridate(V), IrF_6^-

Caesium hexafluoroiridate(V), $CsIrF_6$, is obtained by the action of BrF_3 on a 1:1 mixture of CsCl and $IrBr_3$ (*47, 48*), and was found to have a magnetic moment of 1.29 B.M. at 298 °K, and behaviour clearly indicating a low-spin $^3T_{1g}$ (t_{2g}^4) ground state (*49*). *Brown et al.* (*32*) reported only two bands between 10 and 40 kK. — at 21 and 30 kK. — which they assigned as $^3T_{1g} \rightarrow {}^5E_g$ and $^3T_{1g} \rightarrow {}^3E_g$, $^3T_{2g}$, $^3A_{1g}$, and $^3A_{2g}$, but the spectrum of *Allen et al.* (*10*) revealed relatively weak bands at 12.9, 19.8, and 24.2 kK., as well as a moderately strong absorption at 6.55 kK. and an indication of another near 4 kK. (Fig. 10).

On the assumption of the identity of their 19.8 kK. band with the 21 kK. band of *Brown et al.*, *Allen*, *El-Sharkawy*, and *Warren* rejected the $^3T_{1g} \rightarrow {}^5E_g$ assignment on the basis of comparisons with the isoelectronic Os(IV) species, and attributed all the bands below 20 kK. as arising from transitions within the t_{2g}^4 manifold. Thus the rather indistinct absorption at 12.9 kK. was assigned as the nearly coincident $^3T_{1g} \rightarrow {}^1E_g$ and $^3T_{1g} \rightarrow {}^1T_{2g}$ excitations, and the well defined peak at 19.8 kK. as $^3T_{1g} \rightarrow {}^1A_{1g}$, with the 24.2 kK. absorption attributed to the $^3T_{1g} \rightarrow {}^5E_g$ transition. For a $^3T_{1g}$ (t_{2g}^4) ground state the spin-orbit splitting is of course considerable in the 5*d* series and the absorptions at 6.55 and near 4 kK. were ascribed to the $^3T_{1g}$ (Γ_1) $\rightarrow {}^3T_{1g}$ (Γ_3, Γ_5) and $^3T_{1g}$ (Γ_1) $\rightarrow$ $^3T_{1g}$ (Γ_4) transitions respectively. From these absorptions a good first order estimate of the effective ζ value was obtained.

At higher energies *Allen et al.* observed a well marked moderately strong shoulder at 33.0 kK., and an intense band at 41.7 kK. They agreed with *Brown et al.* in assigning the former band as $^3T_{1g} \rightarrow {}^3E_g$, $^3T_{2g}$, $^3A_{1g}$, and $^3A_{1g}$ (and to $^3T_{1g}$), and fitted their spectrum with parameters $Dq = 2850$ cm^{-1}, $B = 360$ cm^{-1}, and $\zeta = 3400$ cm^{-1}, with $C/B = 4.90$. The 41.7 kK. band could have been attributed to further *d—d* transitions, but on the basis of its considerable intensity and on optical electronegativity grounds, a $\pi \to t_{2g}$ Laporte-allowed excitation was considered more probable.

Table 16. Spectroscopic data for caesium hexafluoroiridate(V)

Band Position (kK.)	Assignment
4.0	$^3T_{1g}\ (\Gamma_1) \to {}^3T_{1g}\ (\Gamma_4)$
6.55	$^3T_{1g}\ (\Gamma_1) \to {}^3T_{1g}\ (\Gamma_5, \Gamma_3)$
12.9	$^3T_{1g}\ (\Gamma_1) \to {}^1T_{2g}, {}^1E_g$
19.8	$^3T_{1g}\ (\Gamma_1) \to {}^1A_{1g}$
24.2	$^3T_{1g}\ (\Gamma_1) \to {}^5E_g$
33.0 (sh)	spin-allowed $t_{2g}^4 \to t_{2g}^3\, e_g$ transitions
41.7	spin-allowed $t_{2g}^4{}^4 \to t_{2g}^3\, e_g$ transitions $+\pi \to t_{2g}$

$Dq = 2850$ cm^{-1}, $B = 360$ cm^{-1}, $\zeta = 3400$ cm^{-1}, $B/C = 4.90$.

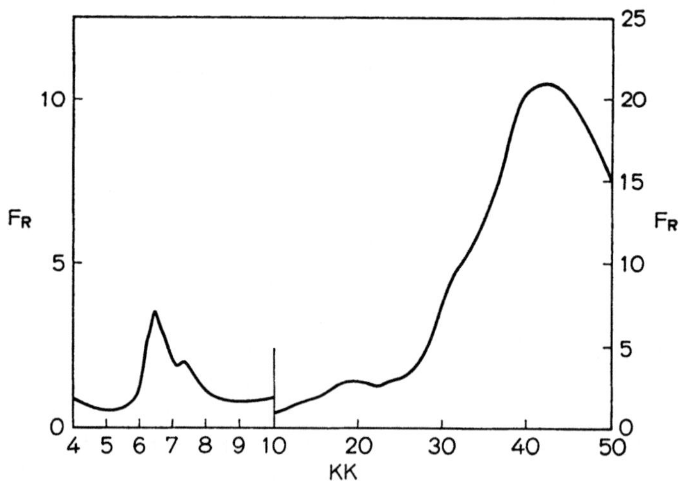

Fig. 10. Electronic spectrum of $CsIrF_6$

e) *Hexafluoroplatinate*(V), PtF_6^-

The hexafluoroplatinate(V) anion was first obtained (*58*) as the dioxygenyl salt, $[O_2^+]\,[PtF_6^-]$, by the action of an oxygen-fluorine mixture on platinum sponge at 450 °C. The potassium salt, $KPtF_6$, was derived from O_2PtF_6 by the action of KF in IF_5 (*59*), and found to have a magnetic moment of 0.87 B.M. at 296 °K, thereby indicating the presence of a low-spin $^2T_{2g}$ (t_{2g}^5) ground state. The electronic spectrum of the O_2^+ salt was studied by *Bartlett* and *Lohmann* by diffuse reflectance and revealed a steeply rising absorption above 20 kK. culminating in a single peak at 28.6 kK. This could possibly represent one or more *d—d* transitions but the authors' description seems more likely ro refer to a $\pi \rightarrow t_{2g}$ charge-transfer transition.

The range covered by these spectral measurements was not indicated but in the lower energy region further weak bands due to the $^2T_{2g} \rightarrow {}^4T_{1g}$ and $^2T_{2g} \rightarrow {}^4T_{2g}$ transitions might be anticipated. Moreover, assuming an effective ζ of ca. 4000 cm^{-1} for Pt(V) a moderately strong band at about 6—8 kK. would be predicted, corresponding to the $^2T_{2g}$ $(\Gamma_7) \rightarrow$ $^2T_{2g}$ (Γ_8) transition.

(iii) MF_6 Systems

The elements of the 5*d* series from Re to Pt all form hexafluorides by direct combination of the elements. They are volatile, strongly oxidising species, highly unstable to moisture, but despite the considerable experimental difficulties electronic spectra have been obtained for these compounds, notably by *Moffitt, Goodman, Fred,* and *Weinstock* (*60*). (Fig. 11). In this work the authors treated the intermediate coupling situation by means of the $t_{2g}{}^n—p^{6-n}$ isomorphism (*61*), using the *jj* coupling scheme, and although their results were expressed in this terminology we have throughout used the more familiar formalism of the strong field, O* double group, approach.

It is also pertinent to note that many of the investigations of the spectra of the 5*d* hexafluorides have concentrated attention on the considerable vibrational structure exhibited by the *d—d* bands, and on the possible consequences of Jahn-Teller instabilities. Since we are here concerned primarily with the electronic energy levels we have not surveyed this vibronic material in detail, but have cited a representative selection. As regards the operation of the Jahn-Teller effect, this is usually manifested more markedly in the vibrational rather than the electronic spectrum. For the $^2T_{2g}$ (t_{2g}^1), $^3T_{1g}$ (t_{2g}^2), $^4A_{2g}$ (t_{2g}^3), and $^3T_{1g}$ (t_{2g}^4) states involved here the lowest lying O* components are respectively Γ_8, Γ_3 (Γ_5), Γ_8, and Γ_1, so that the latter, the t_{2g}^4 system, is Jahn-Teller impotent. For the t_{2g}^3 configuration the Γ_8 state arises from an orbital singlet, and any Jahn-Teller effect will be due to spin only, whilst even for the t_{2g}^1 (Γ_8) and t_{2g}^2 (Γ_3) systems the instability is occasioned only by a t_{2g} orbital degeneracy and would not be expected to yield large static distortions. Consequently, as for the M(IV) and M(V) systems, we have discussed the electronic spectra largely without invoking the operation of the Jahn-Teller effect.

a) *Rhenium hexafluoride*, ReF_6

The magnetic susceptibility of ReF_6 was studied by *Selig et al.* (*62*) between 14 and 296 °K, an effective moment of 0.25 B.M. being reported. In the electronic spectrum *Moffitt et al.* found bands at 5.2 and 32.5 kK., which were assigned as $^2T_{2g}$ $(\Gamma_8) \rightarrow {}^2T_{2g}$ (Γ_7) and $^2T_{2g}$ (Γ_8) ^2E_g (Γ_8) respectively, and deduced the parameters $Dq = 3200$ cm^{-1} and $\zeta = 3500$ cm^{-1}. Subsequently it was shown by *Eisenstein* (*63*) that allowing for configuration interaction between the two Γ_8 states the values $Dq = 3000$ cm^{-1} and $\zeta = 3200$ cm^{-1} were required, and furthermore that the postulation of a small tetragonal Jahn-Teller distortion could lower these to 2950 cm^{-1} and 2500 cm^{-1} respectively. Although the frequencies of the ε_g and τ_{2g} fundamentals (*64*) lend some support to the postulation of a Jahn-Teller distortion a detailed study of the vibronic structure of the 5.2 kK. band by *McDiarmid* (*65*) gave inconclusive results. However, *Brand, Goodman*, and *Weinstock* (*66*) found evidence for a small trigonal distortion.

Above the peak at 32.5 kK. the spectrum of *Moffitt et al.* shows a broad continuing absorption up to 50 kK., but assignment is complicated by the absence of precise intensity data. However, *McDiarmid* (*67*) has examined the high energy region of the ReF_6 spectrum and reports bands at 47.7, 49.3, and 56.4 kK.: the latter two were assigned as members of a *Rydberg* series, and the former as a charge-transfer transition, although its nature was thought uncertain. Nevertheless, optical electronegativities strongly suggest its assignment as a *Laporte*-allowed $\pi \rightarrow t_{2g}$ transition,

Table 17. Spectroscopic data for rhenium hexafluoride

Band Position (kK.)	Assignment
5.0	$^2T_{2g}$ $(\Gamma_8) \rightarrow {}^2T_{2g}$ (Γ_7)
32.5	$^2T_{2g}$ $(\Gamma_8) \rightarrow {}^2E_g$ (Γ_8)
47.7	$\pi \rightarrow t_{2g}$
49.3	?
56.4	?

$Dq = 3000$ cm^{-1}, $\zeta = 3400$ cm^{-1}.

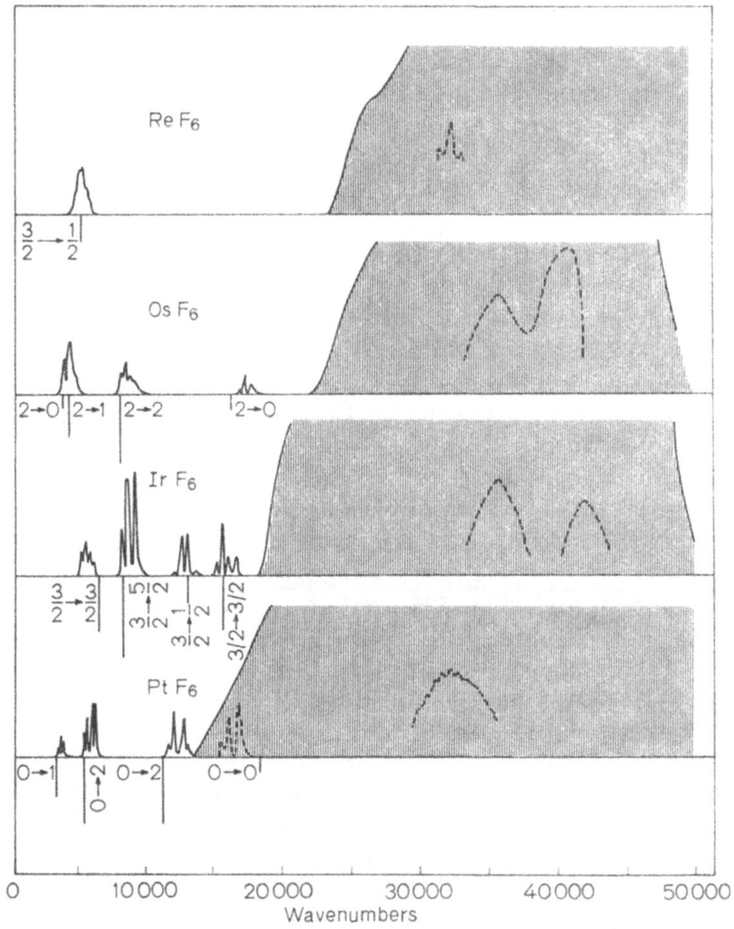

Fig. 11. Electronic spectra of ReF_6, OsF_6, IrF_6, and PtF_6

and its energy is in moderate agreement with the prediction of 44 kK. given by *Warren* (*18*). (This latter was partly based on the 57 kK. reported (*68*) for $\pi \rightarrow t_{2g}$ which it now seems (*67*) may be too low).

b) *Osmium hexafluoride*, OsF_6

The magnetic susceptibility of OsF_6 was measured by *Hargreaves* and *Peacock* (*69*) over the range 81 — 297 °K and a moment of 1.50 B.M. at 297 °K determined. The electronic spectrum due to *Moffitt et al.* shows four bands — at 3.9 and 4.3 kK. (fairly strong) and at 8.5 and 17.3 kK.

(rather weak) — generally agreed to be *d—d* bands, and a broad strong absorption beginning at about 22.5 kK. with maxima at 35.7 and 40.8 kK. whose assignment is less straightforward.

The four lowest energy bands all show considerable vibrational structure and a vibronic analysis has been attempted by *Eisenstein* (*70*), whose assignments we follow in Table 18. On the basis of a diagnonalisation of the complete d^2 O* energy matrices the maxima at 4.32 and 4.37 kK. were ascribed to the $^3T_{1g}$ $(\Gamma_3) \rightarrow {}^3T_{1g}$ (Γ_4) and $^3T_{1g}$ $(\Gamma_3) \rightarrow {}^3T_{1g}$ (Γ_1) transitions, and the 3.9 kK. peak to a hot band, *i.e.* one involving a transition from a vibrationally excited lower state. In addition the peak at 8.48 kK. was assigned to the $^3T_{1g}$ $(\Gamma_3) \rightarrow {}^1T_{2g}$ (Γ_5) and ^1E_g (Γ_3) transitions, and that at 17.3 kK. to the excitation $^3T_{1g}$ $(\Gamma_3) \rightarrow {}^1A_{1g}$ (Γ_1). Note that the stronger 3.9—4.3 kK. band corresponds to a formally spin-allowed transition and the weaker 8.5 and 17.3 kK. bands to spin-forbidden excitations.

Moffitt et al. considered it probable that the bands at 35.7 and 40.8 kK. represented t_{2g}^2—$t_{2g}\,e_g$ *d—d* excitations, but *Eisenstein* (*70*) felt it as likely that these were in fact charge-transfer bands. In an attempt to fit all the bands though *Eisenstein* obtained the parameters Dq — 3450 cm^{-1}, $B = 400$ cm^{-1}, $C = 1505$ cm^{-1}, $\zeta_{55} = 3200$ cm^{-1}, and $\zeta_{35} = 1900$ cm^{-1}, but both ζ_{35} and C/B seem unreasonably low. On the other hand *Moffitt et al.*, fitting only the t_{2g}^2 manifold bands obtained $\zeta = 3400$ cm^{-1} and $(3B + C) = 2400$ cm^{-1}, and these values were found to be satisfactory to fit the spectra of IrF_6 and PtF_6 also.

Moreover, *Jørgensen* (*71*) has argued that the 35.7 and 40.8 kK. bands are more likely to represent charge-transfer transitions, both from electronegativity considerations and on account of their much greater intensity relative to the lower energy transitions. Recently also one of us (*18*) has shown that when relativistic effects are properly included the higher energy spectra of the MF_6 series, and of the MF_6^{2-} and MF_6^- complexes, may all be rationalised on this basis, and we therefore incline to this interpretation. In this context therefore it is noteworthy that in the series OsF_6, IrF_6, and PtF_6, the spectra of *Moffitt et al.* (*60*) clearly show the start of the charge-transfer region moving progressively to lower energies on passing from Os to Ir to Pt. Unfortunately the spectra are incomplete in the higher energy regions, although the authors make clear that the bands above ca. 20—25 kK. do represent much more intense absorptions than those emanating from the t_{2g}^n configurations. Consequently we have attempted to estimate the position of the barycentre of the first Laporte-allowed charge-transfer band from the published spectra, rather than to assume these necessarily to be coincident with the absorption maxima, which do not always correlate with the onset positions of the higher energy regions.

For OsF_6 the absorption appears to increase steadily from near 23 kK. to the first maximum at 35.7 kK. and we therefore take this value for the first $\pi \rightarrow t_{2g}$ excitation: a similar assignment is also adopted for the 40.8 kK. maximum. The optical electronegativity values derived therefrom, and from a similar treatment of the IrF_6 and PtF_6 spectra, are discussed in Section 5 (q.v.).

Table 18. Spectroscopic data for osmium hexafluoride

Band Position (kK.)	Assignment
4.3	$^3T_{1g}\ (\Gamma_3) \rightarrow {}^3T_{1g}\ (\Gamma_4)$
4.4	$^3T_{1g}\ (\Gamma_3) \rightarrow {}^3T_{1g}\ (\Gamma_1)$
8.5	$^3T_{1g}\ (\Gamma_3) \rightarrow {}^1T_{2g}, {}^1E_g$
17.3	$^3T_{1g}\ (\Gamma_3) \rightarrow {}^1A_{1g}$
35.7	$\pi \rightarrow t_{2g}$
40.8	$\pi \rightarrow t_{2g}$

$B = 310\ cm^{-1}$, $\zeta = 3400\ cm^{-1}$, $C/B = 4.75$.

c) *Iridium hexafluoride*, IrF_6

The magnetic susceptibility of IrF_6 was studied by *Figgis, Lewis*, and *Mabbs* (*46*) who found a moment of about 2.90 B.M. at 300 °K. The electronic spectrum was originally studied by *Moffitt et al.*, but a more detailed investigation of the low energy region has recently been carried out by *Brand, Goodman*, and *Weinstock* (*72*).

In the lower energy region *Moffitt et al.* found four bands, all rather weak, at 6.4, 9.0, 13.0, and 16.0 kK. respectively. There is general agreement that these correspond to transitions from the $^4A_{2g}$ ground state within the t_{2g}^3 manifold, and they are all therefore formally spin-forbidden and consequently relatively weak. The more detailed work of *Brand et al.* shows that in the vapour phase the origins of the electronic transitions are located at 6.26, 8.33, 8.86, 12.33 and 15.16 kK., these corresponding to the excitations $^4A_{2g}\ (\Gamma_8) \rightarrow {}^2E_g\ (\Gamma_8)$, $^4A_{2g}\ (\Gamma_8) \rightarrow {}^2T_{1g}\ (\Gamma_8)$, $^4A_{2g}\ (\Gamma_8) \rightarrow {}^2T_{1g}\ (\Gamma_6)$, $^4A_{2g}\ (\Gamma_8)$, $\rightarrow {}^2T_{2g}\ (\Gamma_7)$ and $^4A_{2g}\ (\Gamma_8) \rightarrow {}^2T_{2g}\ (\Gamma_8)$ respectively. These bands are all fitted reasonably well by the parameters suggested by *Moffitt et al.* $\zeta = 3400\ cm^{-1}$, and $(3B + C) = 2400\ cm^{-1}$. As anticipated *Brand et al.* found no evidence of Jahn-Teller instability for the $^4A_{2g}\ (\Gamma_8)$ ground state, although such effects were detected for the excited states $^2E_g\ (\Gamma_8)$ and $^2T_{2g}\ (\Gamma_8)$.

Above about 18 kK. the beginning of a region of intense absorption is evident in the spectrum of *Moffitt et al.* and two maxima at 35.5 and 42.0 kK. were attributed to spin-allowed *d—d* transitions however, for the reasons outlined in Section 3. (iii). c we prefer, with *Jørgensen* (*71*), to assign this region as originating in Laporte-allowed charge-transfer excitations. Furthermore this intense charge-transfer region clearly begins some 5 kK. or so below that of OsF_6 and we therefore take the well marked shoulder at ca. 28 kK. as the first $\pi \rightarrow t_{2g}$ band (*78*) and assign the maxima at 35.5 and 42.0 kK. to further $\pi \rightarrow t_{2g}$ transitions.

Table 19. Spectroscopic data for iridium hexafluoride

Band Position (kK.)	Assignment
6.3	${}^4A_{2g}\,(\Gamma_8) \rightarrow {}^2E_g\,(\Gamma_8)$
8.3	${}^4A_{2g}\,(\Gamma_8) \rightarrow {}^2T_{1g}\,(\Gamma_8)$
8.9	${}^4A_{2g}\,(\Gamma_8) \rightarrow {}^2T_{1g}\,(\Gamma_6)$
12.3	${}^4A_{2g}\,(\Gamma_8) \rightarrow {}^2T_{2g}\,(\Gamma_7)$
15.2	${}^4A_{2g}\,(\Gamma_9) \rightarrow {}^2T_{2g}\,(\Gamma_8)$
28.0 (sh)	$\pi \rightarrow t_{2g}$
35.5	$\pi \rightarrow t_{2g}$
42.0	$\pi \rightarrow t_{2g}$

$B = 305$ cm^{-1}, $\zeta = 3400$ cm^{-1}, $C/B = 4.90$.

d) Platinum hexafluoride, PtF_6

The magnetic susceptibility of PtF_6 shows an almost temperature independent paramagnetism between 170 and 294 °K (*73*) (c.f. the iso-electronic OsF_6^{2-} and IrF_6^- anions), corresponding to a magnetic moment of about 1.40 B.M. at 300 °K. The spectrum of *Moffitt et al.* shows four band systems in the lower energy region which were attributed to *d—d* transitions within the t_{2g}^4 subshell. These were centred at 3.3, 5.5, 12.0, and 16.0 kK. respectively, of which the first was rather weak, the second moderately strong and highly structured, and the third slightly weaker but again with well marked structure. The fourth band was partly obscured by the onset of intense absorptions, but its vibronics were clearly discernible.

No vibrational analysis has been given for PtF_6, but the absorption at 4.81 kK. seems most likely to correspond to a hot band (c.f. OsF_6) and full d^4 O* calculations suggest that the Γ_5 and Γ_3 components of ${}^3T_{1g}$ should not be separated by more than about 200 cm^{-1}. On this basis the 3.34 kK. band may be assigned as $^3T_{1g}$ $(\Gamma_1) \rightarrow$ $^3T_{1g}$ (Γ_4) and the peaks at 5.24 and 5.42 in the spectrum of *Moffitt et al.* seem to be reasonable candidates for the $^3T_{1g}$ $(\Gamma_1) \rightarrow$ $^3T_{1g}$ (Γ_5) and $^3T_{1g}$ $(\Gamma_1) \rightarrow$ $^3T_{1g}$ (Γ_3) transitions. The 12 kK. band system clearly represents the $^3T_{1g}$ $(\Gamma_1) \rightarrow$ $^1T_{2g}$ (Γ_5) and $^3T_{1g}$ $(\Gamma_1) \rightarrow$ ^1E_g (Γ_3) transitions, but since these should not be separated by more than about 150 cm^{-1} the spectrum does not show which peaks should be assigned to these individual excitations. The 16 kK. band is though clearly the $^3T_{1g}$ $(\Gamma_1) \rightarrow$ $^1A_{1g}$ (Γ_1) transition, but it is obscured to such an extent by the high intensity bands as to render nugatory any speculation about the position of its origin. As before the low energy bands are reasonably well fitted by $\zeta = 3400$ cm^{-1} and $(3B + C) = 2400$ cm^{-1}, but for PtF_6, as for the Os and Ir compounds, it is not possible to deduce any reliable value of Dq. It does though seem probable that this parameter is never less than about 3000—3500 cm^{-1}.

As before we incline to the view that the higher energy bands represent charge-transfer transitions. The peak at 32 kK. shows some vibrational structure which might argue for a *d—d* spin-allowed assignment, but its high intensity makes a charge-transfer assignment more reasonable. The published spectrum though clearly implies the presence of a high intensity peak between 17 and 27 kK. and we follow *Jørgensen* (*71, 74*) in estimating its position to be in the region of 25 kK.. This and the 32 kK. band we therefore assign as $\pi \rightarrow t_{2g}$ Laporte-allowed transitions whilst the 45.5 kK. maximum seems more likely to represent a dominantly $\sigma \rightarrow t_{2g}$ excitation. The optical electronegativities derived for Pt(VI) and the other M(VI) fluorides are discussed together in Section 5.

Table 20. Spectroscopic data for platinum hexafluoride

Band Position (kK.)	Assignment
3.3	$^3T_{1g}$ $(\Gamma_1) \rightarrow$ $^3T_{1g}$ (Γ_4)
5.2	$^3T_{1g}$ $(\Gamma_1) \rightarrow$ $^3T_{1g}$ (Γ_5)
5.4	$^3T_{1g}$ $(\Gamma_1) \rightarrow$ $^3T_{1g}$ (Γ_3)
12.0	$^3T_{1g}$ $(\Gamma_1) \rightarrow$ $^1T_{2g}, {}^1E_g$
16.0	$^3T_{1g}$ $(\Gamma_1) \rightarrow$ $^1A_{1g}$
25 (est)	$\pi \rightarrow t_{2g}$
32.0	$\pi \rightarrow t_{2g}$
45.5	$\sigma \rightarrow t_{2g}$ (?)

$B = 300$ cm^{-1}, $\zeta = 3400$ cm^{-1}, $C/B = 5.0$.

4. Nephelauxetic Effects and Related Considerations

(i) General Considerations

In our previous survey (*1*) of the hexafluoro complexes of the $3d$ series we discussed in some detail the information which may be derived from a consideration of the value of the nephelauxetic ratio, $\beta\,(=B_{\text{complex}}/B_{\text{gas}})$. To obtain this parameter it is clearly necessary to have available values of the free-ion *Racah* parameter, B, and for the $4d$ elements these may be obtained either from the compilation of *Di Sipio et al.* (*75*) or by use of the *Jørgensen-Racah* relationship (*76, 77*).

$$B = 472 + 28q + 50(z+1) - 500/(z+1)$$

where z is the formal ionic charge and q the occupation number of the d^q shell. (C.f. the similar relationship used (*1*) for the $3d$ series). As before we have preferred to adopt this latter expression which yields results in good agreement with *Jørgensen's* (*78*) empirical suggestion, $B, M'^{n+}(4d^q) = 0.66\,B, M^{n+}(3d^q)$, and have assumed that the C/B ratios obtained by *Tanabe* and *Sugano* (*25*) for elements of the $3d$ series also hold for the corresponding elements and oxidation states of the $4d$ (and $5d$) series. For the $5d$ series there are insufficient experimental free-ion data available for the Slater-Condon F_k parameters (and hence B and C) to have been derived, and we have therefore adopted *Jørgensen's* relationship, $B, M''^{n+}(5d^q) = 0.60\,B, M^{n+}(3d^q)$. In Table 21 we list the values

Table 21. Free-ion values of the *Racah* parameter, B, for $4d$ and $5d$ elements

$4d$ Series: Element	M^0	M^+	M^{2+}	M^{3+}	M^{4+}	M^{5+}	C/B ‡
Mo	190	462	567	631	687	717	4.4
Tc	218	490	595	659	706	745	4.6
Ru	246	518	623	687	734	773	4.75
Rh	274	546	651	715	762	801	4.90
Pd	302	574	679	743	790	829	5.0
Ag	330	602	707	771	818	857	5.0

Note: The Slater-Condon parameters of *Di Sipio et al.* (*75*), derived from experimental data, lead to B values in reasonable agreement with those listed above for M^+ and M^{2+} ions, but give appreciably larger values for M^0. Fitting of the data of Table 22 on this basis would lead to appreciably smaller values of z_{eff} than the listed values, especially where these are relatively low.

‡ For M^{3+}, M^{4+}, and M^{5+}.

Table 21 (continued)

5*d* Series: Element*	M^0	M^+	M^{2+}	M^{3+}	M^{4+}	M^{5+}	M^{6+}
Re	224	426	520	586	642	692	740
Os	259	461	554	621	677	727	775
Ir	294	496	589	656	712	762	809
Pt	329	530	624	691	746	797	844

* For C/B ratios use same values as for corresponding 4*d* element.

Note: Where the 5*d* values are derived from 3*d* results based on empirical Slater-Condon parameters, the same caveat applies as for the 4*d* elements. [C.f. also Ref. (*1*)].

thereby calculated for various oxidation states of the 4*d* and 5*d* elements, the C/B ratios employed also being indicated.

From a knowledge of the free-ion B values the β parameters are readily derived. For most d^q configurations these represent β_{35} quantities, but as explained before (*1*) for d^3 and d^8 systems respectively β_{55} and β_{33} parameters may also in principle be deduced. In the present work the β values listed are always β_{35} parameters unless otherwise stated, but for the $5d^4$ systems OsF_6^{2-} and IrF_6^- and the $5d^2$ system OsF_6 the β values are essentially β_{55} quantities since they are obtained predominantly from bands arising from the t_{2g}^n manifold. The β and Dq parameters derived for the 4*d* and 5*d* hexafluorides are now given in Table 22.

Essentially there are two covalent contributions which may be operating within the nephelauxetic effect. In the first place there is central field covalency due to the reduction of the effective positive charge on the cation by the screening of the d^q configuration by the ligands, and secondly symmetry restricted covalency due to participation of the metal e_g and t_{2g} orbitals in molecular orbital formation with appropriate symmetry adapted ligand combinations. Unfortunately it is not a simple matter to assess, from β values alone, the relative magnitudes of the central field and symmetry restricted contributions. Thus *Jørgensen* (*78*) has shown that there should exist an approximate proportionality between B_{gas} and $(z+Z)$, where z is the ionic charge and Z a small constant, but in a complex the corresponding quantity becomes a^4 $(z_{eff}+Z)$, where a is the *Stevens'* (*79*) delocalisation coefficient, and z_{eff} the effective cationic charge. Consequently both contributions are involved, the central field effect by the reduction of z to z_{eff}, and symmetry restricted covalency by the intervention of the a^4 term.

However one can investigate the two extreme possibilities — all central field covalency, or all symmetry restricted covalency. The former limit is obtained either by setting $a^4=1$ and finding z_{eff} from some empirical expression of the form $B_{gas} =$ constant. $(z_{eff} + Z)$, or more simply by solution of the *Jørgensen-Racah* equation for z, knowing B_{obsd}, and the latter limit by assuming $z_{eff} = z$, so that the coefficient of presence of the metal orbital, a^2, is given to a good approximation simply by $\beta^{\frac{1}{2}}$. It is though possible to obtain a more reasonable solution to the problem by following *Jørgensen's* (*78*) suggestion that the two effects be assumed to be of equal importance, and thus to assign $\beta^{\frac{1}{4}}$ to each. We then obtain $a^2 = \beta^{\frac{1}{4}}$, and derive z_{eff} using $B = B_{gas}.\ \beta^{\frac{1}{4}}$, and our earlier treatment of the $3d$ hexafluoro anions supports the view that this constitutes a not unreasonable compromise. In Table 22 therefore we treat the available $4d$ and $5d$ data as described above and list both the extreme results, z_{min} and $a^2{}_{min}$, and the equal weighting results z_{root} and $a^2{}_{root}$. For comparison the results for the first transition series are summarised in Table 23: some of these data are taken from our previous compilation (*1*), but for the MF_6^{3-} complexes more recent results are now to hand (*4*).

In the $5d$ series however it is possible to derive additional information bearing upon the problem of the relative extent of central field and symmetry restricted covalency. For many $5d$ complexes reasonable estimates of the effective spin-orbit coupling constant can be derived from the spectra, and thence the relativistic ratio, β^* $(= \zeta$ complex$/\zeta_{gas})$. When both β and β^* are known for a given system, *Jørgensen* (*74*) has suggested how estimates of both covalency contributions may be made.

Thus for a given complex the repulsion parameter, B, is proportional to a^4 $(z_{eff} + Z)$, and the spin-orbit coupling constant to a^2 $(z_{eff} + Z)$ ^{2}, where z_{eff} and Z are as defined previously. Alternatively we may write B as proportional to a^4 Z^*, where Z^* is a measure of $\langle r \rangle^{-1}$, so that $\beta = a^4$ $[Z^*(\text{complex})/Z^*(\text{gas})]$, and similarly for proportional to $a^2 Z^{*2}$ we obtain $\beta^* = a^2$ $[Z^*(\text{complex})/Z^*(\text{gas})]$, as long as ζ_{ligand} is small with respect to ζ_{metal}, as is of course the case for the $5d$ elements. Consequently, knowing β and β^* both a^2 and Z^* (complex)$/Z^*$(gas) (written as Z/Z_0) may be determined, and are listed in Table 22 for $5d$ compounds. As may be seen from the above derivation (Z/Z_0) is not a direct measure of the relative effective charges in complex and free-ion, but does provide a measure of the extent to which central field covalency influences the B and ζ parameters.

The derivation of both β and β^* is of course subject to some uncertainty. In neither case are the free-ion values (of B and ζ) well established, and, in addition to *Jørgensen*'s approximation for B ($5d$), we have adopted the values of ζ deduced by *Dunn* (*80*) for various $5d$ ions, amplified where necessary by extra values obtained via the known proportionality (*81*) between B^3 and ζ. The free-ion values of ζ used are listed in Table 24.

We have though previously expressed our reservations (*10, 11*) concerning the combination of β and β^* data for the derivation of a^2 and (Z/Z_0). Thus, whereas the repulsion parameter, B, is essentially an outer radial quantity (*82, 83*) the spin-orbit coupling constant, ζ, is dominantly an inner orbital function (*84*). Moreover the (Z/Z_0) values derived in most cases indicate a rather small measure of central field covalency. Nevertheless, the a^2 values obtained tend to parallel the a^2_{min} rather than the a^2_{root} values, especially as the extent of covalency increases from the M(IV) to M(V) to M(VI) series, thus suggesting that a^2_{root} values may be too large. On the other hand the a^2_{min} values are likely to be too small since comparison of ligand weighting coefficients for Ir(IV) species derived from the spectra (*11*) with those deduced by measurements (*85, 86*) reveals the former to indicate significantly greater ligand involvement. Nonetheless it seems evident that as the overall extent of covalency (as measured by β and β^*) increases the importance of the symmetry restricted contribution similarly grows and in fact becomes predominant. Support for this interpretation may also be derived from the theoretical study of the distance dependence of ζ due to *Al-Mobarak* and *Warren* (*84*), who showed that more than 90% of the total value of ζ_{3d} arose from the region within 1 a.u. of the nucleus (see Fig. 12) so that ζ should be little affected by the action of central field effects on the outer radial regions of the d-orbital. For the $5d$ series no suitable wave functions were available, but ζ_{5d} would be expected to be an even more inner radial function, thus leading to a similar conclusion. Consequently symmetry restricted covalency would be predicted to play the major role and to become dominant for low values of β.

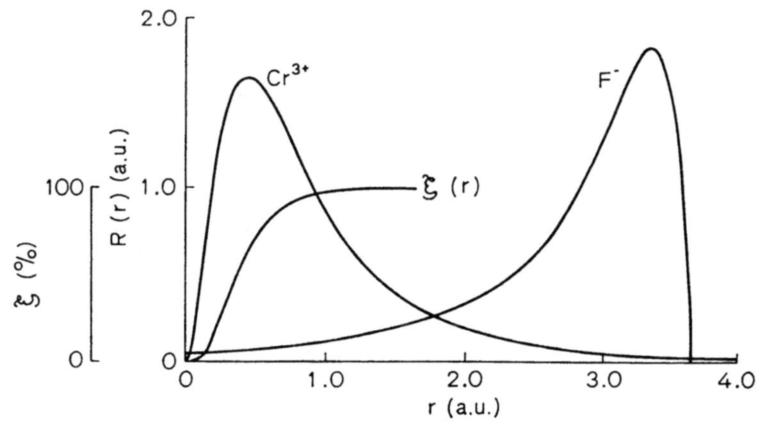

Fig. 12. Distance dependence of the spin-orbit coupling constant

Table 22. Interpretations of the nephelauxetic effect

Complex	B	ζ	β	β^*	z_{min}	a^2_{min}	z_{root}	a^2_{root}	$a^2(\zeta)$	Z/Z_0 (ζ)
4*d* Complexes:										
MoF_6^{3-}	570		0.90		2.05	0.95	2.35	0.97		
RuF_6^{3-}	550		0.80		1.25	0.89	1.9	0.95		
RhF_6^{3-}	460		0.64		0.6	0.80	1.2	0.89		
AgF_6^{3-}	472		0.61		0.4	0.78	1.0	0.88		
TcF_6^{2-}	530		0.75		1.35	0.87	2.25	0.93		
RuF_6^{2-}	500		0.68		0.9	0.82	1.8	0.91		
RhF_6^{2-}	410		0.54		0.4	0.73	1.1	0.86		
PdF_6^{2-}	400		0.43		0.1	0.66	0.7	0.81		
TcF_6^{-}	520		0.70		1.2	0.84	2.4	0.91		
RuF_6^{-}	425		0.55		0.6	0.74	1.45	0.86		
5*d* Complexes:										
ReF_6^{2-}	543[a]	2250	0.85	0.77	2.3	0.92	2.95	0.96	0.98	0.93
OsF_6^{2-}	500[a]	2900	0.74	0.75	1.35	0.86	2.4	0.93	0.90	0.91
IrF_6^{2-}	510	3300	0.72	0.73	1.1	0.85	2.2	0.92	0.90	0.90
PtF_6^{2-}	380	—	0.53	—	0.3	0.73	1.15	0.85	—	—
OsF_6^{-}	410[a]	3200	0.56	0.71	0.55	0.75	1.9	0.87	0.77	0.97
IrF_6^{-}	360[a]	3400	0.47	0.66	0.1	0.69	1.3	0.83	0.70	0.97
OsF_6	310[a]	3400	0.40	0.64	0.1	0.63	1.25	0.80	0.63	0.99
IrF_6	305[a]	3400	0.38	0.57	0	0.62	1.0	0.79	0.63	0.95
PtF_6	300[a]	3400	0.36	0.47	−0.1	0.60	0.7	0.77	0.65	0.85

[a] Values essentially B_{55} quantities.

Table 23. Interpretations of the nephelauxetic effect in the 3*d* series

Complex	B_{35}	β_{35}	z_{min}	a^2_{min}	z_{root}	a^2_{root}
VF_6^{3-} [a]	670	0.78	1.5	0.88	2.1	0.94
CrF_6^{3-} [a]	735	0.80	1.5	0.89	2.1	0.95
MnF_6^{3-} [a]	782	0.80	1.4	0.89	2.05	0.95
FeF_6^{3-} [a]	800	0.77	1.2	0.88	1.9	0.94
CoF_6^{3-} [a]	765	0.70	0.7	0.84	1.55	0.91
NiF_6^{3-} [a]	703	0.61	0.3	0.78	1.1	0.88
CuF_6^{3-} [a]	641	0.53	−0.1	0.73	0.7	0.85
CrF_6^{2-}	608	0.57	0.7	0.75	1.65	0.87
MnF_6^{2-}	585	0.55	0.55	0.74	1.5	0.86
CoF_6^{2-}	635	0.53	0.35	0.73	1.2	0.85
NiF_6^{2-}	515	0.41	−0.1	0.64	0.65	0.80

[a]) Data from Ref. (*4*), other values from Ref. (*1*).

Table 24. Free-ion spin-orbit coupling constant values

Ion	ζ (cm^{-1})	Ion	ζ (cm^{-1})	Ion	ζ (cm^{-1})
Re(IV)	3300	Re(V)	4000	Re(VI)	4800
Os(IV)	3870	Os(V)	4500	Os(VI)	5265
Ir(IV)	4500	Ir(V)	5180	Ir(VI)	6000
Pt(IV)	5100	Pt(V)	5950	Pt(VI)	7250

(ii) Hexafluorometallate Systems

In this Section we survey the nephelauxetic and related parameters for the various 4*d* and 5*d* hexafluoro species, and where appropriate make comparisons with existing data for 3*d* complexes and for other hexahalo systems. As far as the *Dq* parameter is concerned the results show no very marked dependence on oxidation state in the 4*d* and 5*d* series, but values for 5*d* elements are always somewhat greater than those for the corresponding elements of the same oxidation state in the 4*d* block. Throughout the results though the nephelauxetic ratio, β, is found (for a given oxidation state) to decrease towards the end of a series, thereby reflecting the increasing tendency towards covalency. Similarly the β values generally increase on passing from the 3*d* through the 4*d* to the 5*d* series, for a given valency, thus underlining the increasing stability of the higher oxidation states between the first and the third transition series.

In the 4*d* and 5*d* series the z_{root} values calculated as described above show a similar spread to that found earlier (*1*) for the 3*d* complexes. Thus of the nineteen species listed, only two show z_{eff} as less than +1, and only six a z_{eff} greater than +2: the remaining anions all show z_{eff} between +1 and +2, which *Jørgensen* suggested (*76*) was the normative result. It is also interesting to examine the z_{eff} values for various oxidation states of a given element; with the exception of Tc (and the parameters for TcF_6^- are somewhat uncertain) the calculated z_{eff} value actually falls on passing progressively to higher positive oxidation states, *e.g.* Ru(III), 1.9; Ru(IV), 1.8; Ru(V), 1.45 and Os(IV), 2.4; Os(V), 1.9; Os(VI), 1.25. This of course closely parallels the behaviour previously observed for the 3*d* series.

(iii) Comparison with other Hexahalo Systems

In general the stability of hexahalo complexes of the transition metals tends to decrease appreciably with increasing atomic number and with increasing size of the halide ligand. (See Ref. (*1*) for discussion of the 3*d* situation.) Thus for example in the 4*d* series MF_6^{3-} species are known for Mo, Ru, Rh, and Ag, but MCl_6^{3-} ions have been reported only for the first three of these metals. Similarly in the 5*d* series MF_6^- anions exist for all the elements from Ta to Au, but MCl_6^- species have been reported only as far as WCl_6^-, whilst MF_6 compounds extend from W to Pt, but MCl_6 systems only up to $ReCl_6$.

In comparing the spectroscopic parameters for hexafluorides with other halide data we have here confined our attention largely to hexachlorides. This is for two reasons. In the first place it is for the chlorides that the most useful data are available since for bromides and iodides the charge-transfer bands occur at lower energies and may obscure some or all of the *d—d* excitations. Secondly, for both F^- and Cl^- ζ_p, the spin-orbit coupling constant of the ligand, is quite small (ca. 220 and 550 cm^{-1} respectively) with respect to the values obtaining for the metals, especially those of the 5*d* block, whereas for Br^- and I^- ζ_p values of ca. 2200 and 6000 cm^{-1} are appropriate, thus complicating the estimation of the effective spin-orbit coupling constant of the metal from the value of the relativistic ratio, β^*.

The *Dq* and *B* values reported for hexachloro species of the 4*d* and 5*d* series therefore given in Table 25, together with those of various derived quantities. In the Table the effective metal ζ values are also given, where these are available. Unfortunately, for the hexachloro anions of Ru(III) and Pd(IV) no *d–d* spectra are to hand, whilst for those of Ru(IV) and Rh(IV) they are obscured by the onset of the charge-transfer region. Where data are listed they tend to follow the same pattern as found for the 3*d* elements: both *Dq* and β values are generally smaller than for the corresponding fluoro compounds, thereby underlining the greater extent of covalency anticipated. In keeping with this the calculated z_{root} quantities are all appreciably lower than for the hexafluorides, but again follow the tendency noted by *Jørgensen* (*76*) to lie between +1 and +2.

Finally, it is of interest to compare the estimates of covalency contributions for Ir(IV) hexahalides deduced by *Allen et al.* (*11*) from spectroscopic data, with those obtained by *Owen* and *Thornley* (*85, 86*) from ESR results. These latter authors attributed the reduction of ζ below the free-ion value, entirely to symmetry restricted covalency, deriving the expression $\zeta_{obsd} = N_\pi^2\,(\zeta_d + \frac{1}{2}\,\alpha_\pi^2\zeta_p)$, where the normalising constant, N_π, is equal to $(1-4\alpha_\pi S + \alpha_\pi^2)^{-\frac{1}{2}}$, and ζ_d and ζ_p are the metal and ligand spin-orbit coupling constants respectively. Assuming $S \sim o$ the fractional contributions of the metal and ligands are equal to $(1 + \alpha_\pi^2)^{-1}$ (which we write as α_π^2) and $\frac{1}{2}\alpha_\pi^2\,(1 + \alpha_\pi^2)^{-1}$ respectively. Since the ligand t_{2g} symmetry orbitals contain equal contributions from four ligand atoms, the latter quantity formally corresponds to 4 f_π, where f_π is the fraction of unpaired electron spin density per ligand deduced from the ESR *g* values (or ligand hyperfine splittings).

For IrF_6^{2-}, $IrCl_6^{2-}$, and $IrBr_6^{2-}$ the f_π values derived from the *g* values were 11.5, 14.1, and 14.1% respectively, whilst the hyperfine splittings gave 8.1, 8.0, and 7.5%. The spectroscopic ζ_{obsd} values of *Allen et al.* (*11*) yield f_π values of 6.9, 10.3, and 14.8% respectively, which are clearly of comparable magnitude and these probably constitute as reasonable an estimate of the ligand involvement as do *Thornley's* two somewhat disparate results. Nevertheless, although the results of *Al-Mobarak* and *Warren* (*84*) do suggest that symmetry restricted covalency makes the dominant contribution to ζ_{obsd}, these values are likely somewhat to exaggerate the ligand contributions because of the neglect of the central field effects, and it is noteworthy that the a^2 (ζ) values derived from the combination of the β and β^* expressions are appreciably larger than the a_π^2 figures. For convenience we collect in Table 26 the f_π values deduced from a_{min}^2, a_{root}^2, a_π^2, and a^2 (ζ), and from the two ESR approaches. We are inclined to feel that the *g* value and a_π^2 results probably over-estimate the covalency effects, but that they are somewhat underestimated by a_{root}^2. However, irrespective of the actual numerical values, the spectroscopic results clearly indicate the parallel between the decreasing value of ζ_{obsd} and the increasing extent of covalency on changing the ligand from F^- to Cl^- to Br^-.

Table 25. Interpretations of the nephelauxetic effect for hexachloro anions

Complex	Dq	B	β	z_{min}	a^2_{min}	z_{root}	a^2_{root}	ζ	β^*	a^2	Z/Z_0
4*d* Complexes:											
$MoCl_6^{3-}$	1920	440	0.70	0.85	0.84	1.6	0.91				
$RhCl_6^{3-}$	2030	350	0.49	0.15	0.70	0.7	0.84				
$TcCl_6^{2-}$	2500	460	0.65	0.8	0.81	1.7	0.90				
5*d* Complexes:											
$ReCl_6^{2-}$	2900	440	0.68	1.1	0.82	2.1	0.91	2400	0.73	0.86	0.92
$OsCl_6^{2-}$	2500	450	0.66	0.9	0.81	1.95	0.90	2650	0.68	0.87	0.88
$IrCl_6^{2-}$	2300	425	0.60	0.5	0.77	1.55	0.88	2750	0.61	0.84	0.85

Table 26. Ligand participation in Ir(IV) complexes

	IrF_6^{2-}	$IrCl_6^{2-}$	$IrBr_6^{2-}$
a^2_{min}	0.85	0.77	—
f_π	3.8	5.8	—
a^2_{root}	0.92	0.88	—
f_π	2.0	3.0	—
a^2 (ζ)	0.90	0.84	—
f_π	2.5	4.0	—
a^2_π	0.73	0.59	0.41
f_π	6.9	10.3	14.8
$f_\pi(g)$	11.5	14.1	14.1
f_π (hyp.)	8.1	8.0	7.5

5. Charge-Transfer Bands and Optical Electronegativities

In Fig. 13 is shown the molecular orbital scheme commonly adopted for the description of octahedral transition metal hexahalo complexes. This scheme is too familiar to merit extended discussion and we refer the reader to our previous survey for fuller details (*1*). For any formally d^n complex the occupied levels in the ground state correspond to the dominantly ligand levels up to and including t_{1g}, plus n electrons contained in the predominantly metal $2t_{2g}$ and $2e_g$ levels. For the hexafluoro species of the $4d$ and $5d$ series, unlike certain $3d$ anions, the ground states are always low-spin systems by virtue of the greater Dq and smaller B values prevailing. Consequently the $2t_{2g}$ level is always filled to the maximum extent so that for $n \leqslant 6$ we are dealing with t_{2g}^n ground states.

As shown in Fig. 13 the ligand σ-orbitals give rise to the symmetry adapted combinations $a_{1g}+e_g+t_{1u}$, and the π-orbitals to the levels $t_{1g}+t_{2g}+t_{1u}+t_{2u}$, the latter group lying appreciably higher than the former. In the general case therefore the lowest energy charge-transfer transitions will correspond to $\pi \rightarrow t_{2g}$ (γ_5) excitations, and the highest energy bands to $\sigma \rightarrow e_g$ (γ_3) transitions. The relative ordering of the $\pi \rightarrow e_g$ (γ_3) and $\rightarrow t_{2g}$ (γ_5) excitations will clearly depend essentially on the respective separations of the π and σ, and of the e_g and t_{2g} levels.

For the $4d$ and $5d$ hexafluoro systems the lowest energy charge-transfer transitions are thus $\pi \rightarrow \gamma_5$ type excitations, except for the t_{2g}^6 ions, RhF_6^{3-}, PdF_6^{2-}, and PtF_6^{2-}, and the $t_{2g}^6\, e_g^2$ anion, AgF_6^{3-}, for which the lowest energy bands are all $\pi \rightarrow \gamma_3$. In Table 27 therefore we list the positions of the charge-transfer bands found for the hexafluoro systems of the $4d$ and $5d$ series, together with their assignments. Data for the $3d$ ions have previously been listed for both MF_6^{2-} (*1*) and MF_6^{3-} (*4*) ions.

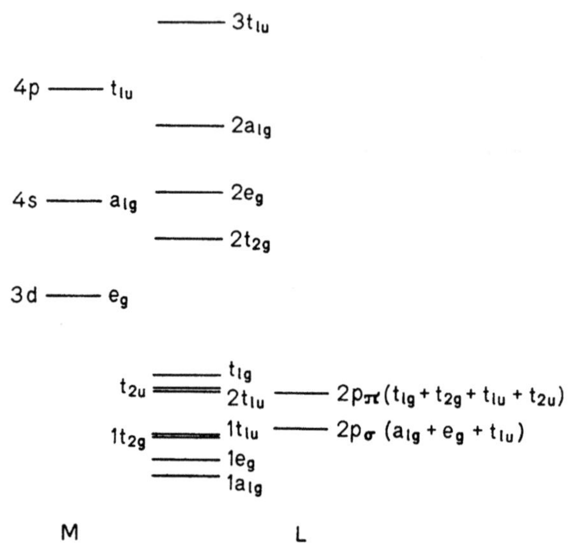

Fig. 13. Molecular orbital scheme for octahedral fluoride complexes

Charge-transfer excitations from odd ligand levels to the even metal γ_5 and γ_3 levels clearly represent formally Laporte-allowed $u \to g$ transitions, and consequently should be intense. Ligand to metal transitions involving even ligand orbitals are of course also possible, but would be parity forbidden and are therefore rather seldom observed. For many of the ions here treated though the data are derived from reflectance measurements and the intensity criterion is of limited value because of the increase in the scattering coefficient which usually occurs above about 25 kK. [c.f. (*1*)].

Nevertheless, our present understanding of charge-transfer transitions in hexahalo species owes much to the early solution measurements of *Jørgensen* (*24*) on hexachloro, hexabromo-, and hexaiodo- anions of the $4d$ and $5d$ series in which many charge-transfer assignments were put forward largely on the basis of the observed intensities. *Jørgensen* (*87*) has recently reinforced the arguments for these assignments and latterly further support for the charge-transfer attributions has come from MCD studies on, for example, Os(IV) and Ir(IV) systems (*88, 89*), which also shed light on the ordering of the predominantly ligand levels.

On the basis of his early work *Jørgensen* (*24, 90*) postulated that in the $4d$ and $5d$ series the positions of the charge-transfer bands in hexahalo systems could be represented by the expressions;

$$(\pi \rightarrow \gamma_5) = V + kD + q\,(A - E)$$

and
$$(\pi \rightarrow \gamma_3) = V + k'D + q\,(A - E)$$

for complexes formally possessing a d^q configuration, and *Allen, El-Sharkawy*, and *Warren* (*4*) showed that these relationships were also valid for hexafluorides of the $3d$ series. Here V is a parameter depending on the nature of the ligand, the oxidation state of the metal, and the particular transition series, and D is the spin-pairing energy. E contains the variation of the orbital energy of the t_{2g} (or e_g) electrons with the occupation number, q, and A is approximately equal to the corresponding Racah parameter. For a metal ion configuration with total spin, S, the spin-pairing energy contribution is $-S(S+1)D$: the values of k and k' for the various d^q systems are thus readily obtained and have been listed in full by *Allen et al.* (*4*).

As will be appreciated in due course, it is necessary explicitly to consider the effects of spin-orbit coupling in dealing with the charge-transfer transitions of complexes of the $5d$ elements. Consequently we defer consideration of this series and treat now only the MF_6^{3-} and MF_6^{2-} species of the $4d$ elements. For the former anions the actual band positions are known only for the RuF_6^{2-} and RhF_6^{2-} anions (Table 27), but for the Tc and Pd complexes values may be estimated by adding 26 kK. to the figures for the corresponding MCl_6^{2-} species (see Table 27.). Of the MF_6^{3-} ions the positions of the charge-transfer bands have been measured only for RuF_6^{3-} and AgF_6^{3-}, but for MoF_6^{3-} and RhF_6^{3-} values may be estimated as before. For the MF_6^- species however too little reliable data is available for further consideration.

For the MF_6^{2-} species a good fit to the experimental values was afforded by the parameters $V = 63$, $D = 4$, and $(E - A) = 5$ kK. (*12*). [c.f. $V = 48$, $D = 4.5$, and $(E - A) = 5.5$ kK. for the $3d$ MF_6^{2-} ions (*1*)], but for the MF_6^{3-} complexes only a moderate agreement between theory and experiment was obtained using $V = 92$, $D = 4.5$, and $(E - A) = 9$ kK., as compared with the values $V = 60$, $D = 6$, and $(E - A) = 5.5$ for the $3d$ MF_6^{3-} ions (*4*). Both the V and the $(E - A)$ values for the $4d$ MF_6^{3-} series seem rather too high, and this largely results from the accommodation of the rather low energy $\pi \rightarrow \gamma_3$ bands in AgF_6^{3-}. It seems possible that this is because the $(E - A)$ quantities relate to the e_g as well as to the t_{2g} orbitals, and that the former are somewhat greater in magnitude. [C.f. also CuF_6^{3-} in the $3d$ series (*4*)]. In any case it seems necessary to assign the first charge-transfer band of AgF_6^{3-} at 27.5 kK. as $\pi_g \to e_g$ and that at 37.5 kK. as $\pi_u \to e_g$ on the basis of the calculated results, which are shown in Table 27, together with the experimental values, for the hexafluoro species of the $4d$ series. If the data for AgF_6^{3-} are omitted the MF_6^{3-} values may be fitted with the parameters $V = 82$, $D = 4$, and $(E-A) =$ 6 kK., which seems not unreasonable in comparison with the $3d$ results.

Table 27. Charge-transfer transitions of $4d$ and $5d$ hexafluoro complexes

MF_6^{3-}	Complex	Band Obsd. (kK.)	Assignment	Calcd. (see Table 30) (kK.)
$4d$ Systems				
	MoF_6^{3-}	(71)*	$\pi \to \gamma_5$	73
	RuF_6^{3-}	55	$\pi \to \gamma_5$	51
	RhF_6^{3-}	(65)*	$\pi \to \gamma_3$	57
	AgF_6^{3-}	27.5, 37.5	$\pi \to \gamma_3$	43
MF_6^{2-}	TcF_6^{2-}	(55)*	$\pi \to \gamma_5$	56
	RuF_6^{2-}	48.0	$\pi \to \gamma_5$	48
	RhF_6^{2-}	39.6, 44.6	$\pi \to \gamma_5$	41
	PdF_6^{2-}	(55)*	$\pi \to \gamma_3$	56
MF_6^-	TcF_6^-	32.0	$\pi \to \gamma_5$ (?)	—
	RuF_6^-	40.0, 50	$\pi \to \gamma_5$	—
$5d$ Systems				
MF_6^{2-}	ReF_6^{2-}	(57)*	$\pi \to \gamma_5$	57
	OsF_6^{2-}	55 (53)*	$\pi \to \gamma_5$	56
	IrF_6^{2-}	47.8	$\pi \to \gamma_5$	45
	PtF_6^{2-}	(64)*	$\pi \to \gamma_3$	67
MF_6^-	ReF_6^-	40	$\pi \to \gamma_5$	49
	OsF_6^-	41.7	$\pi \to \gamma_5$	42
	IrF_6^-	41.7	$\pi \to \gamma_5$	41
	PtF_6^-	28.5	$\pi \to \gamma_5$	31
MF_6	WF_6	57	$\pi \to \gamma_5$	55
	ReF_6	47.7	$\pi \to \gamma_5$	44
	OsF_6	35.7	$\pi \to \gamma_5$	38
	IrF_6	28.0	$\pi \to \gamma_5$	28
	PtF_6	25	$\pi \to \gamma_5$	26

* Values estimated from known band positions in corresponding hexachloro species.

For a given series the value of D, the spin-pairing energy, does not in fact stay constant, but its average value for a given dq configuration is $7/6\ (5/2\,B + C)$, and for MF_6^{3-} ions D ranges only between 3.9 and 4.7 kK.

Similarly for MF_6^{2-} ions D varies between the extremes of 3.5 and 4.2 kK., (*12*) so that in neither case is any sensible error involved in assuming constancy. However, the determination of D and the assignment of the Laporte-allowed charge-transfer bands also permits the evaluation of *Jørgensen's* (*91*) optical electronegativity parameters. These are obtained using the relationship

$$\sigma_{corr} = (\varkappa_{opt}(X) - \varkappa_{opt}(M^{n+})) \times 30 \text{ kK.}$$

where $\varkappa_{opt}(X)$ is the optical electronegativity of the ligand and $\varkappa_{opt}(M^{n+})$ that of the metal cation corresponding to the oxidation state of the complex. For F^- $\varkappa_{opt}(X)$ is 3.9 on the Pauling scale. The value of σ_{corr} is obtained from the position of the lowest Laporte-allowed band, applying two corrections. Firstly $\pi \to \gamma_3$ and $\pi \to \gamma_5$ transitions are compared directly by subtracting Δ (10 Dq) from the former, and secondly allowance is made for the change in the spin-pairing energy accompanying the $d^q \to d^{q+1}$ transition. The derivation of these latter corrections has been described before (*1*), and the quantities have also been explicitly listed.

The optical electronegativity of a given M^{n+} state in fact assesses the ease (or difficulty) with which an electron is transferred from the ligand to the metal. Thus metal oxidation states which readily accept an electron from the ligands show charge-transfer bands at low energies and exhibit high $\varkappa_{opt}$ values. [See also (*1*)]. Of course the nephelauxetic ratio, β, also essentially measures a similar characteristic — the extent of covalency — and is also a measure of electron accession from the ligands towards the central metal atom, as reflected in the reduction of β below unity. Thus one would anticipate a parallel between $\varkappa_{opt}$ and β — high $\varkappa_{opt}$ values being associated with low β values — and a theoretical justification for this on a semi-quantitative level has been given by *Allen* and *Warren* (*92*). Since $\varkappa_{opt}$ is in principle independent of the ligand a correlation should therefore exist when β values are considered for complexes of a constant halide.

Furthermore, it has been shown by *Jørgensen* (*93*) that an additional correlation should exist between $\varkappa_{opt}$ and the d-orbital occupation number, q, and that the slope of the $\varkappa_{opt}$ vs. q plot should approximate to the quantity $(E-A)$. For the $3d$ hexafluoro anions the most recently determined data (*1, 4*) certainly bear out this expectation, both for the MF_6^{3-} and the MF_6^{2-} series. (See Table 28.). For the $4d$ elements the MF_6^{2-} anions again show a good parallel between $\varkappa_{opt}$ and q, and a similar trend is found for the MF_6^{3-} species. For the latter complexes though the plot is less good, probably due to the anomalous behaviour of the d^8 AgF_6^{3-} anion, as discussed above. In all cases though the slopes of the $\varkappa_{opt}$ vs. q

plots are consistent with their interpretation as $(E-A)$, bearing in mind the values of this quantity derived by fitting the band positions according to Eqs. 5(1) and 5(2). For the $4d$ series the plots of $\varkappa_{opt}$ vs. q and $\varkappa_{opt}$ vs. β respectively are shown in Fig. 14 and 15.

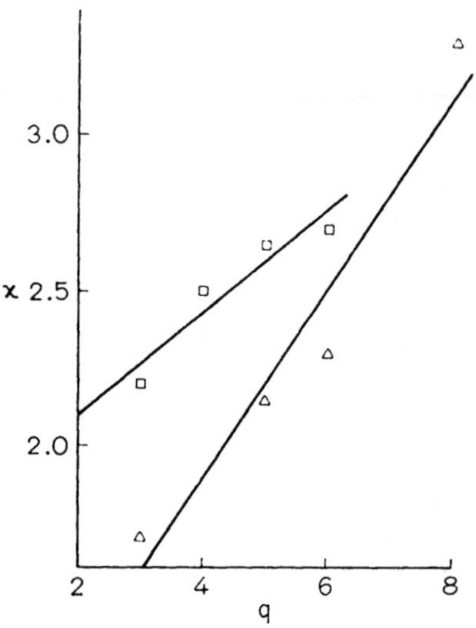

Fig. 14. Correlation between $\varkappa_{opt}$ and q for $4d$ oxidation states

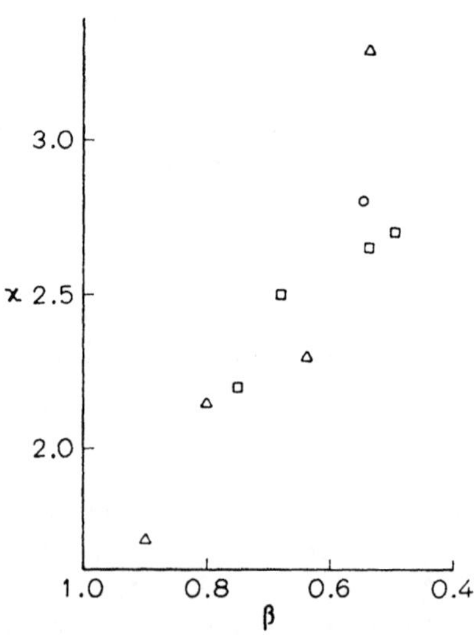

Fig. 15. Correlation between $\varkappa_{opt}$ and β for $4d$ hexafluoro compounds

Table 28. Optical electronegativities for $3d$ and $4d$ oxidation states

	Ti	V	Cr	Mn	Fe	Co	Ni	Cu
$3d$ Systems								
M(III)	2.2	2.25	2.4	2.65	2.9	3.0	3.05	3.25
M(IV)	—	2.6	2.65	3.05	—	3.1	3.4	—
			Mo	Tc	Ru	Rh	Pd	Ag
$4d$ Systems								
M(III)			1.7	—	2.15	2.3	—	3.3
M(IV)			1.9	2.2	2.5	2.65	2.7	—
M(V)			2.05	2.4	2.8	—	—	—

We have not as yet however treated the charge-transfer data available for complexes of the $5d$ series. For these latter species though the effective spin-orbit coupling constants are often of the order of 3 kK. or more, as compared with only about 1 kK. for $4d$ systems, and smaller values still for the $3d$ elements. Consequently, as for the d—d transitions it is often necessary explicitly to consider relativistic effects in the interpretation of charge-transfer spectra, and in particular to make allowance for the changes in spin-orbit contributions which may accompany a given $d^q \rightarrow d^{q+1}$ transition. In fact one of us has shown (*18*) that these changes are indeed comparable to the corrections arising from similar variations in the spin-pairing energy contributions, and must therefore be taken into account in the calculation of optical electronegativities and in the fitting of the $\pi \rightarrow \gamma_5$ (γ_3) bands for the $5d$ hexafluoro series.

Of the various t_{2g}^n ground states the d^0 ($^1A_{1g}$), d^3 ($^4A_{2g}$), and d^6 ($^1A_{1g}$) configurations lead to orbital singlets, unsplit by spin-orbit coupling which thus makes no first order contribution to the energy. For the other t_{2g}^n configurations in the $5d$ series though the familiar strong field formalism is not altogether appropriate for evaluating the ζ changes associated with the $d^q \rightarrow d^{q+1}$ transitions because of the numerous off-diagonal matrix elements connecting the ground state components with higher levels, and it is therefore better to adopt the jj coupling scheme in which the interaction matrices are diagonal in ζ. This is facilitated by use of the t_{2g}^n—p^{6-n} isomorphism (*61*) and the ζ contributions to the ground state Γ components are shown in Table 29 in addition to the spin-orbit corrections to σ_{obsd} thereby necessitated for the calculation of $\varkappa_{opt}$ (*18*). (For comparison the corresponding spin-pairing corrections are also listed). The adoption of the jj coupling scheme may in fact be shown to be justified in a number of typical $5d$ systems in which a ground state purity of greater than 80% is thus obtained. (*18, 24*).

In the $5d$ series the $\varkappa_{opt}$ values calculated with allowance for spin-orbit effects are henceforth denoted by $\overset{*}{\varkappa}_{opt}$, and the simple symbol $\varkappa_{opt}$ used for those values deduced using only the spin-pairing energy corrections. From Table 29 it is seen that the spin-pairing and spin-orbit corrections act in a similar sense for the d^1, d^4, and d^5 configurations but are opposed for the d^2 and d^3 cases: consequently the most noticeable effects of the relativistic corrections should be found for these latter cases. Note also that it is imperative to include the spin-orbit terms for fitting of the band positions according to Eqs. 5 (*1*) and 5 (*2*); these corrections will here naturally be of opposite sign to those listed for the calculation of $\overset{*}{\varkappa}_{opt}$.

In this way it was shown that the $\overset{*}{\varkappa}_{opt}$ values derived from data for MCl_6^{2-}, MF_6^-, and MF_6 species gave excellent correlations with the occupation number, q, and that the $\pi \rightarrow \gamma_5$ (γ_3) peak positions could be well reproduced using Eqs. 5 (*1*) and 5 (*2*), with spin-orbit corrections. In all cases the correlations were significantly better when the relativistic terms were included than when they were omitted, and in Table 30 we list the $\varkappa_{opt}$ and $\overset{*}{\varkappa}_{opt}$ values derived from the $5d$ data for MF_6^{2-}, MF_6^-, and MF_6 complexes. In the Table we also show the observed and the calculated band positions using the corrected forms of Eqs. 5 (*1*) and 5 (*2*). Once again the $\overset{*}{\varkappa}_{opt}$ vs. q plots yield slopes in excellent agreement with the (E—A) values deduced from these equations. Finally, in Figs. 16

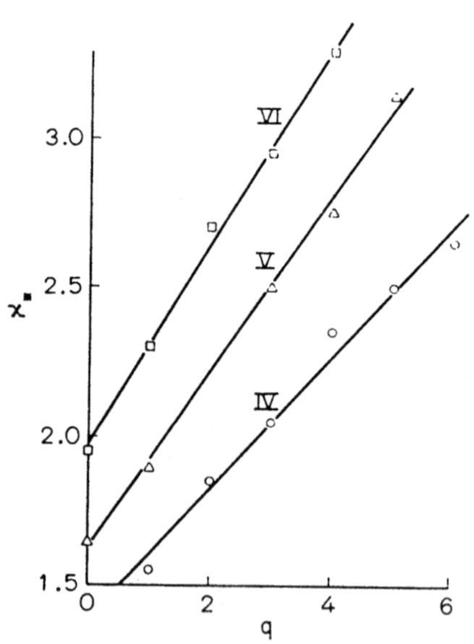

Fig. 16. Correlation between $\overset{*}{\varkappa}_{opt}$ and q for $5d$ oxidation states

and 17 we show that plots of $\varkappa^*_{opt}$ vs. q and of $\varkappa^*_{opt}$ vs. β respectively for the $5d$ series. In the latter the strong correlation between the optical electronegativity and the nephelauxetic ratio already observed in the $3d$ and $4d$ series is again in evidence, thereby providing additional support for our interpretations.

For the $5d$ series therefore the inclusion of the spin-orbit corrections leads to improved $\varkappa_{opt}$ vs. q plots, satisfactory predictions of the charge-transfer band via Eqs 5 (*1*) and 5 (*2*), and a strong correlation between

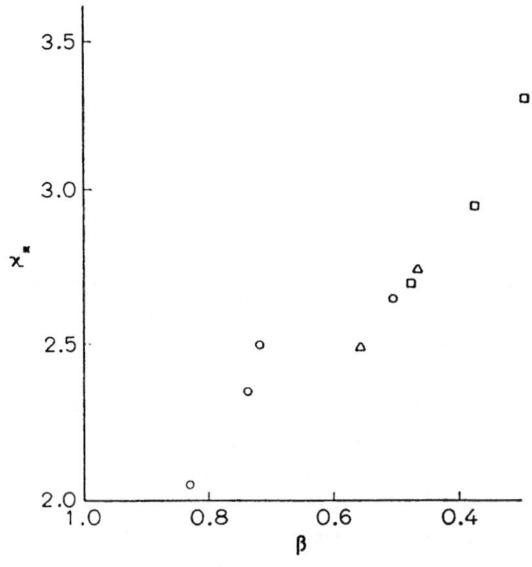

Fig. 17. Correlation between $\varkappa^*_{opt}$ and β for $5d$ hexafluoro compounds

Table 29. Spin-orbit contributions and corrections for use in treatment of charge-transfer data

Ground state contributions							
Occupation number, q	0	1	2	3	4	5	6
Spin-orbit contribution	0	$-\frac{1}{2}\zeta$	$-\zeta$	0	-2ζ	$-\zeta$	0
Corrections to σ_{obsd} for calculation of $\varkappa^*_{opt}$							
Occupation number, q	0	1	2	3	4	5	6
Spin-pairing (D)	0	$+^2/_3$	$+^4/_3$	-2	$-^4/_3$	$-^2/_3$	$+1(+\Delta)$
Spin-orbit (ζ)	$+^1/_2$	$+^1/_2$	-1	$+2$	-1	-1	0

Table 30. Optical electronegativities and fitting parameters for 5*d* systems

		W	Re	Os	Ir	Pt
Optical electronegativities						
M(IV)	$\varkappa_{opt}$	1.75	2.2	2.25	2.4	2.65
	$\overset{*}{\varkappa}_{opt}$	1.85	2.05	2.35	2.5	2.65
M(V)	$\varkappa_{opt}$	1.95	2.45	2.75	2.65	3.0
	$\overset{*}{\varkappa}_{opt}$	1.9	2.5	2.5	2.75	3.15
M(VI)	$\varkappa_{opt}$	2.0	2.25	2.6	3.15	3.2
	$\overset{*}{\varkappa}_{opt}$	1.95	2.15	2.7	2.95	3.3

Fitting parameters (kK.)

	V	D	$(E—A)$	ζ	$\overset{*}{\varkappa}_{opt}$ vs. q slope
MF_6^{2-}	79	4.5	8	3	6.6
MF_6^-	68	3	8.5	3	8.9
MF_6	58	2.5	9.5	3	10.0

$\overset{*}{\varkappa}_{opt}$ and β. In particular the rather anomalous behaviour of d^3 systems such as Re(IV), Os(V), and Ir(VI) in the $\varkappa_{opt}$ vs. q sequence is rectified. In Table 30 we also list the V, D, $(E—A)$, parameters used to fit the $\pi \rightarrow \gamma_5$ (γ_3) bands via Eqs. 5 (*1*) and 5 (*2*) and the slopes of the $\overset{*}{\varkappa}_{opt}$ vs. q plots. Once again the values of D and ζ will not be strictly constant throughout any given series, but the errors introduced thereby will not exceed about 1 kK.. Having regard to the uncertainties attending the determination of the B, C, and ζ quantities in the 5*d* series a more precise evaluation of the D and ζ terms is probably not justified.

The overall trends in optical electronegativity values throughout the three transition series are thus clear. In all cases $\varkappa_{opt}$ increases from left to right for a given oxidation state within a given series, thereby reflecting increasing covalency and decreasing stability, especially for the higher oxidation states. Similarly $\varkappa_{opt}$ ($\overset{*}{\varkappa}_{opt}$) decreases on passing from the 3*d* to the 4*d* to the 5*d* series for any given oxidation state of the same orbital occupation number, q, this time corresponding to the well known increase in stability of the higher oxidation states on passing from the first to the second to the third transition series.

6. Conclusions

The present work in conjunction with our earlier (*1*) treatment of the 3*d* data constitutes a tolerably complete survey of the electronic spectra of all the transition metal hexafluoro complexes. A number of obvious gaps still remain — for example the neutral hexafluorides (MF_6) of the $4d$ series — and more data for the MF_6^- $5d$ series would also be welcome, as would further study of some of the systems here treated. The general trends are however clear and in this respect the study of fluoro species has proved of great value by virtue of the ability of the fluoride ligand to stabilise such a wide range of oxidation states, the higher of which are frequently not accessible in other systems. Some aspects of the interpretation of the spectra still present difficulties, for example the separation of central field and symmetry restricted covalency effects, and in this matter the results of photoelectron spectroscopy and ESCA studies may well prove valuable. Thus, for the MCl_6^{2-} anions of the $5d$ series ESCA studies (*94*) have established that the effective positive charge on the metal decreases from left to right across the series, just as suggested by our interpretation of the spectroscopic data in Table 25. There is however a substantial discrepancy in the magnitudes of the metal positive charges thereby deduced, which clearly renders desirable a reconsideration of the meaning and significance of the z_{eff} parameters customarily used to describe such situations, and further work is in progress directed towards this end.

7. References

1. *Allen, G. C., Warren, K. D.:* Struct. Bonding *9*, 49 (1971).
2. *Dutta-Ahmed, A., Boudreaux, E. A.:* Inorg. Chem. *12*, 1590 (1973).
3. — — Inorg. Chem. *12*, 1597 (1973).
4. *Allen, G. C., El-Sharkawy, G. A. M., Warren, K. D.:* Inorg. Chem. *10*, 2538 (1971).
5. *Demuynck, J., Veillard, A., Wahlgren, U.:* J. Am. Chem. Soc. *95*, 5563 (1973).
6. *van der Lugt, W. Th. A. M.:* Intern. J. Quantum Chem. *6*, 859 (1972).
7. — Chem. Phys. Letters *10*, 117 (1971).
8. *Biloen, P., Prins, R.:* Chem. Phys. Letters *16*, 611 (1972).
9. *Allen, G. C., Warren, K. D.:* Inorg. Chem. *8*, 1895 (1969).
10. — *El-Sharkawy, G. A. M., Warren, K. D.:* Inorg. Chem. *11*, 51 (1972).
11. — *Al-Mobarak, R., El-Sharkawy, G. A. M., Warren, K. D.:* Inorg. Chem. *11*, 787 (1972).
12. — *El-Sharkawy, G. A. M., Warren, K. D.:* Inorg. Chem. *12*, 2231 (1973).
13. *Liehr, A. D., Ballhausen, C. J.:* Ann. Phys. (N.Y.) *6*, 134 (1959).
14. *Eisenstein, J. C.:* J. Chem. Phys. *34*, 1628 (1961).
15. — J. Chem. Phys. *35*, 2246 (1961).
16. *Schroeder, K. A.:* J. Chem. Phys. *37*, 2553 (1962).
17. — J. Chem. Phys., *37*, 1587 (1962).
18. *Warren, K. D.:* J. Inorg. Nucl. Chem., in press.
19. *Peacock, R. D.:* Progr. Inorg. Chem. *2*, 193 (1960).
20. *Toth, L. M., Brunton, G. D., Smith, G. P.:* Inorg. Chem. *8*, 2694 (1969).
21. *Peacock, R. D.:* Chem. Ind. (London) *1956*, 1391.

22. — J. Chem. Soc. *1955*, 3291.
23. *Schmidtke, H. H.:* Z. Physik. Chem. (Frankfurt) *40*, 96 (1964).
24. *Jørgensen, C. K.:* Mol. Phys. *2*, 309 (1959).
25. *Tanabe, Y., Sugano, S.:* J. Phys. Soc. Japan *9*, 753, 766 (1954).
26. *Hoppe, R., Homann, R.:* Naturwissenschaften *53*, 501 (1966).
27. *Edwards, A. J., Peacock, R. D.:* Chem. Ind. (London) *1960*, 1441.
28. *Schwochau, K., Herr, W.:* Angew. Chem. *75*, 95 (1963).
29. *Jørgensen, C. K., Schwochau, K.:* Z. Naturforsch. A *20*, 65 (1965).
30. *Hepworth, M. A., Peacock, R. D., Robinson, P. L.:* J. Chem. Soc. *1954*, 1197.
31. *Figgis, B. N., Lewis, J., Nyholm, R. S., Peacock, R. D.:* Disc. Faraday Soc. *26*, 103 (1958).
32. *Brown, D. H., Russell, D. R., Sharpe, D. W. A.:* J. Chem. Soc. A *1966*, 18.
33. *Sharpe, A. G.:* J. Chem. Soc. *1950*, 3444.
34. *Nyholm, R. S., Sharpe, A. G.:* J. Chem. Soc. *1952*, 3579.
35. *Sharpe, A. G.:* J. Chem. Soc. *1953*, 197.
36. *Hoppe, R., Klemm, W.:* Z. Anorg. Allgem. Chem. *268*, 364 (1952).
37. *Allen, G. C., Warren, K. D.:* Inorg. Chem. *8*, 753 (1969).
38. *Hargreaves, G. B., Peacock, R. D.:* J. Chem. Soc. *1957*, 4212.
39. — — J. Chem. Soc. *1958*, 3776.
40. *Walton, R. A., Crouch, P. C., Brisdon, B. J.:* Spectrochim. Acta A *24*, 601 (1968).
41. *Hugill, D., Peacock, R. D.:* J. Chem. Soc. A *1966*, 1339.
42. *Tanner, K. N., Duncan, A. B. F.:* J. Am. Chem. Soc. *73*, 1164 (1951).
43. *Weinstock, B., Claassen, H. H., Chernick, C. L.:* J. Chem. Phys. *38*, 1470 (1963).
44. *Westland, G. J.:* quoted in Ref. 19 (Ph. D. thesis).
45. *Peacock, R. D.:* J. Chem. Soc. *1956*, 1291.
46. *Figgis, B. N., Lewis, J., Mabbs, F. E.:* J. Chem. Soc. *1961*, 3138.
47. *Hepworth, M. A., Robinson, P. L., Westland, G. J.:* J. Chem. Soc. *1954*, 4269.
48. — — — J. Chem. Soc. *1958*, 611.
49. *Earnshaw, A., Figgis, B. N., Lewis, J., Peacock, R. D.:* J. Chem. Soc. *1961*, 3132.
50. *Dorain, P. B., Paterson, H. H., Jordan, P. C.:* J. Chem. Phys. *49*, 3845 (1968).
51. *Dickinson, J. R., Johnson, K. E.:* Mol. Phys. *19*, 19 (1970).
52. *Douglas, I. N.:* J. Chem. Phys. *51*, 3066 (1969).
53. *Wheeler, T. E., Perros, T. P., Naeser, C. P.:* J. Am. Chem. Soc. *77*, 3488 (1955).
54. *Jørgensen, C. K.:* Acta. Chem. Scand. *12*, 1539 (1958).
55. *Leary, K., Bartlett, N.:* Chem. Commun. *1972*, 903.
56. *Walton, R. A., Crouch, P. C., Brisdon, B. J.:* Spectrochim. Acta. *24A*, 601 (1968).
57. *Peacock, R. D.:* J. Chem. Soc. *1957*, 467.
58. *Bartlett, N., Lohmann, D. H.:* J. Chem. Soc. *1962*, 5253.
59. — — J. Chem. Soc. *1964*, 619.
60. *Moffitt, W., Goodman, G. L., Fred, M., Weinstock, B.:* Mol. Phys. *2*, 109 (1959).
61. *Griffith, J. S.:* The theory of transition metal ions. London: Cambridge University Press 1964.
62. *Selig, H., Cafasso, F. A., Gruen, D. M., Malm, J. G.:* J. Chem. Phys. *36*, 3440 (1962).
63. *Eisenstein, J. C.:* J. Chem. Phys. *33*, 1530 (1960).
64. *Levin, I. W., Abramowitz, S., Muller, A.:* J. Mol. Spectry. *41*, 415 (1972).
65. *McDiarmid, R.:* J. Mol. Spectry. *38*, 495 (1971).
66. *Brand, J. C. D., Goodman, G. L., Weinstock, W.:* J. Mol. Spectry. *38*, 449 (1971).

67. *McDiarmid, R.:* J. Mol. Spectry. *39*, 332 (1971).
68. *Tanner, K. N., Duncan, A. B. F.:* J. Am. Chem. Soc. *73*, 1164 (1951).
69. *Hargreaves, G. B., Peacock, R. D.:* Proc. Chem. Soc. *1959*, 85.
70. *Eisenstein, J. C.:* J. Chem. Phys. *34*, 310 (1961).
71. *Jørgensen, C. K.:* Mol. Phys. *3*, 201 (1960).
72. *Brand, J. C. D., Goodman, G. L., Weinstock, B.:* J. Mol. Spectry. *37*, 464 (1971).
73. *Zupan, J., Pirkmajer, E., Slivnik, J.:* Croat. Chem. Acta. *39*, 135 (1967).
74. *Jørgensen, C. K.:* Struct. Bonding *1*, 3 (1966).
75. *Di Sipio, L., Tondello, E., De Michaelis, G., Oleari, L.:* Inorg. Chem. *9*, 927, (1970).
76. *Jørgensen, C. K.:* Helv. Chim. Acta Fasciculus Extraordinarius Alfred Werner *1967*, 131.
77. *Racah, G.:* Lunds Univ. Arsskr. *50*, No. 21, 31 (1954).
78. *Jørgensen, C. K.:* Progr. Inorg. Chem. *4*, 73 (1962).
79. *Stevens, K. W. H.:* Proc. Roy. Soc. (London) *219A*, 542 (1953).
80. *Dunn, T. M.:* quoted in Ref. 46 and 49.
81. *Cole, G. M., Garrett, B. B.:* Inorg. Chem. *9*, 1898 (1970).
82. *Ranson, G. S. S., Warren, K. D.:* Chem. Phys. Letters *10*, 236 (1971).
83. *Brown, D. A., Fitzpatrick, N. J.:* J. Chem. Soc. A *1966*, 941.
84. *Al-Mobarak, R., Warren, K. D.:* Chem. Phys. Letters *21*, 513 (1973).
85. *Owen, J., Thornley, J. H. M.:* Rep. Progr. Phys. *29*, 675 (1966).
86. *Thornley, J. H. M.:* Proc. Phys. Soc. (London) (Solid State Physics) [2] *1*, 1024 (1968).
87. *Jørgensen, C. K.:* Progr. Inorg. Chem. *12*, 101 (1970).
88. *Schatz, P. N., Shiflett, R. B., Spencer, J. A., McCaffery, A. J., Piepho, S. B., Dickinson, J. R., Lester, T. E.:* Symp. Faraday Soc. *3*, 14 (1969).
89. *Piepho, S. B., Lester, T. E., McCaffery, A. J., Dickinson, J. R., Schatz, P. N.:* Mol. Phys. *19*, 781 (1970).
90. *Jørgensen, C. K.:* Absorption spectra and chemical bonding in complexes. London: Pergamon Press 1962.
91. — Orbitals in atoms and molecules. New York: Academic Press 1962.
92. *Allen, G. C., Warren, K. D.:* Mol. Phys. *20*, 379 (1971).
93. *Jørgensen, C. K.:* Mol. Phys. *6*, 43 (1963).
94. *Cox, L. E., Hercules, D. M.:* J. Electron Spectr. *1*, 193 (1972/3).

Received March 5, 1974

Structure and Bonding: Index Volume 1-19

GPSR Compliance

The European Union's (EU) General Product Safety Regulation (GPSR) is a set of rules that requires consumer products to be safe and our obligations to ensure this.

If you have any concerns about our products, you can contact us on ProductSafety@springernature.com

In case Publisher is established outside the EU, the EU authorized representative is:

Springer Nature Customer Service Center GmbH
Europaplatz 3
69115 Heidelberg, Germany

Batch number: 10620141

Printed by Printforce, the Netherlands